The Mercury Method of Chart Comparison

By

LOIS M. RODDEN

Secretary
Professional
Astrologers
Incorporated

PAI APPROVED TEXT No. 6

AMERICAN FEDERATION OF ASTROLOGERS, INC.
P O BOX 22040
TEMPE, ARIZONA 85282

ISBN No. 0-86690-150-7
Second Printing 1981

Printed in the United States of America
AMERICAN FEDERATION OF ASTROLOGERS, INC.
BOX 22040, TEMPE, ARIZONA 85282

*Dedicated to my daughters
Amy, Lynn and Betts in
appreciation for their
encouragement; and to my
sons Dana and Jonathan
for their patience.*

P.A.I. APPROVED TEXTS

No. 1	Time Changes in the U.S.A.
No. 2	Time Changes in Canada-Mexico
No. 3	Time Changes in the World
No. 4	Astrology: 30 Years Research
No. 5	Horoscopes of U.S. Presidents
No. 6	The Mercury Method of Chart Comparison

In whatever way you come to me,
in that way
shall I appear to you.

from the *Bhavagad Gita*

Table of Contents

CHAPTER I

Premise

CHART COMPARISON, or Synastry, is the study of the relationship between two people, by comparing the planetary positions of one person's natal and progressed horoscope to the planetary positions of the other person's natal and progressed horoscope.

The premise of THE MERCURY METHOD OF CHART COMPARISON is that relationships can be defined with consistent accuracy by using the planet Mercury to compare charts. This book deals exclusively with the aspects that Mercury in one person's chart makes to the planets in another person's chart.

Mercury opens the gates between two people. Mercury shows a clear picture of both the attitude and the circumstances between two people. Mercury the messenger carries awareness of one person to the other. Wherever Mercury of one chart aspects a planet of another chart, there is the area of communication and understanding.

The nature of the aspect and of the planet aspected indicates the harmony or discord, love or hate, gain or loss, endurance or separation between two people.

Why do we relate to one person and not another? Why does one person stimulate our curiosity, our attraction, our good will, and another person stimulate our animosity, our challenge, or our indifference? Why do we have deep enduring bonds with one person and casual inter-action with another?

Life is filled with people who could make contact but pass each other by in the minor events of every day. The wonder is not in our lack of contact but in the miracle that we *do* understand each other. We do need each other. We do communicate.

We are each the sum total of all we have thought, felt and done. Our thought process, consciously or unconsciously, is the sum total of all that has come before. Out of this well, or chalice or awareness, we make choices, we act, and we interact with others.

All experience stems from thought. Thought is the basis of the universe, the origin of creation and the doorway of expression. It is not only the instigator of motive but the release mechanism of action.

Plato said, "Thoughts rule the world."

In astrology, Mercury is thought. It is the cerebral activator, the definitive evaluator, the storehouse and dispenser of data. Mercury weighs and measures, discerns and defines, draws comparisons and differences to judge and determine. This is the process which supports the growth of discrimination.

In mythology, Mercury was the messenger of the gods, sexless and neutral in nature, who carried the awareness, or level of communication, from one point to another.

Mercury is the planet that carries the awareness, or level of communication from *one person to another.*

In the Mercury Method of Chart Comparison, the delineation is read from Mercury's point of view, as this is his thought-awareness of the other person. The person receiving the aspect has his own reality, that may or may not coincide. However, Mercury's action that stems from his attitude will emphasize or alleviate a commensurate or opposing attitude from the other person.

Our choice in relationship is determined by what and who we are, by how we think. The measure of relationship is in correspondence of aspects that Mercury in one person's chart makes to the planets in the other person's chart.

CHAPTER II

The Mercury Method

THE NECESSARY REQUISITES for comparing charts are: 1. A knowledge of the symbols of the planets and of their nature. 2. A knowledge of the sequence of zodiacal signs, in order to determine the aspect (distance) between the two planets. 3. A knowledge of where to read this information in an Ephemeris.

Anyone who can read the planets and aspects can use the Mercury Method of Chart Comparison. A skilled astrologer will work with two accurately drawn natal charts, progressed to the year of interest. By considering all factors, individually and comparatively, he draws a more detailed and comprehensive report.

The preliminary to chart comparison is the consideration of two separate people. The individual must not be lost in the technicalities at any point in astrology. The statement that the sign Aries is harmonious with the signs Leo and Sagittarius is technically true, but may *not* be true to every individual Aries. Each person is a complex of astrological factors as well as personal background. To read the chart, consider the person.

Consider the age and sex, the environment and history. What is the person's educational, economic, and emotional status? Where are his interests *now*? What does he seek in this relationship? What is he prepared to give to this relationship?

Consider what measure of growth he has made by his past experience and understanding. A person who has qualities and conditions that are astrologically discordant may well have learned through his obstacles to have a positive attitude. No Mars aspect can give violence to a person who does not have violence as a potential in his psychology as well as in his chart.

Jupiter can only give benefit to the extent that it is individually indicated, nor does fortune often drop into our lap without some type of effort. No Venus aspect can affect the person who has tightly closed his heart to any expression of yielding warmth. The person who has cultivated emotional serenity in trust is open to loving relationships.

Consider the level of *how* the indications between two individuals can apply. Love is possible between any two people, but it will be a different level of love between an old lady and her young grandson, between competitive co-workers, between two friends, or between a virile man and a receptive woman.

Financial benefit is possible between any two people, but it will be on a different level between two people who have corporate business dealings as compared to two people on the tight budget of new career or schooling.

The second preliminary to the Mercury Method is this standard procedure: 1. Consider the sign and aspect harmony of Sun, Moon and Ascendant in the two charts. 2. Place the planets of one chart into the houses of the other chart, to judge one person's influence on the departments of life of the other person. 3. Consider all conjunctions and oppositions with a five degree orb to read in the manner that natal aspects are read.

Where the Mercury Method makes no pretention of being the exclusive or final study in comparing charts, Mercury's role in chart comparison is a consistent indicator in the relationships between people.

The Mercury study is based on the research of 520 charts, with a total catalogue of 5800 aspects. The study includes for the first time a definition of the effect of two planets in north parallel, two planets in south parallel, and two planets in a split parallel (when one planet is in north declination and the other planet is in south declination in the same degree).

After the preliminary steps have been considered, the Mercury Method is followed by this procedure:

With the two natal charts on hand, draw an aspect table, as shown by the example of Mary and John Smith on page 6.

In column 1, Mary's natal Mercury is recorded by longitude and declination. In examining John's natal chart, we find that Mary's Mercury is making a 6° opposition aspect to John's Mercury. This is recorded in column 2.

In John's natal chart, Saturn is at 13° Aries. Mary's Mercury is in 1° square aspect to John's Saturn. This is recorded in column 2. Mary's Mercury is in a split parallel aspect to John's Mercury. This is recorded in column 2. This is the total of aspects that Mary's Mercury makes to John's natal chart.

In column 3, record John's natal Mercury by longitude and declination. It repeats the Mercury aspects to Mercury as recorded for Mary. John's Mercury is inconjunct Mary's natal Mars, 1°, and semi-sextile Mary's Neptune, 1°. Record this in column 4.

The year of interest to Mary and John was 1968. Examine the progressed planetary positions of both charts. Record Mary's progressed Mercury by longitude and declination in column 1. Mary's Mercury at 18° Cancer made only one aspect to John's natal and progressed chart, the opposition to Mercury. This is recorded in column 2. Mary's *natal* Mercury was making two aspects to John's 1968 progressed chart; the trine to John's Sun and the semi-sextile to John's Mars. Record in column 2.

In 1968, John's progressed Mercury is recorded in column 3, by declination and longitude. John's Mercury progressed makes one aspect to Mary's natal and progressed chart, a north declination parallel to her natal Mars. John's *natal* Mercury makes a south declination parallel to Mary's Venus progressed. Record the two aspects in column 4.

To delineate the Mercury Method of Chart Comparison between Mary and John, read the Definition of Aspect, chapter 3; read Mercury in aspect to Mercury, chapter 4; and the delineation of Mercury opposition Mercury, Mercury split-parallel Mercury in that chapter. Read each planetary aspect in this order, by aspect definition, planetary chapter, and specific planetary aspect delineation.

The first and most meaningful aspect between two person's charts is the Mercury relationship to Mercury. In 92% of the example charts of relationship, there was an aspect of Mercury

Aspect Table

	Aspects made by Mary to John		Aspects made by John to Mary
☿ Natal 12 ♋ 15 8 N.21	Split P ☿ ☍ ☿ 6° □ ♄ 1°	Natal 18 ♉ 05 ℞ 8 S.50	Split P ☿ ☍ ☿ 6° ⚻ ♂ 2° ⚺ ♆ 1°
Progressed 1968 ☿ 18 ♋ 00 ℞ 22 N.08	☍ ☿	Progressed 1968 2 ♓ 15 0 N.21	N.P ♂
Natal ☿	△ ☉p ⚺ ♂p	Natal ☿	S.P ♀p

to Mercury, natally or progressed to the time of interaction. The nature of that aspect determined whether the communication came with ease or difficulty, with opportunity, with agitation or worry, with spontaneity, or through compelling circumstances.

Mercury in aspect to Mercury is the contact point.

Mercury will make both harmonious and discordant aspects. There is certainly a measure of ambivalence in deep relationships. Moreover, the squares and oppositions are stimulating and productive, more so than the 'easy' aspects. Excessive harmony in the individual chart depletes initiative. Excessive harmony in compared charts is a bore.

To judge duration of the relationship, the first point to consider is the natal strength of aspects. When there are six or more, or even four strong natal Mercury aspects on both charts, the relationship can continue through periods of years when there are only one or two Mercury aspects by progression. The closer that Mercury and the planet are to the exact degree of aspect, the stronger the bond and impact of the two people.

When the natal aspects between the two charts are few, association is maintained only during the period of time that the progressed Mercury aspects are strong. On the other hand, we do not look for an enduring association when both natal and progressed Mercury aspects between the two charts are few or weak.

The larger influence, or control, is in the hands of the person whose Mercury makes the larger number of aspects to the other chart. It is he who makes the strongest impression on the other.

The nature of the association is denoted first of all by the natal chart comparison. The influence of a planet not in natal Mercury aspect may develop with a progression, but if in neither natal nor progressed aspect, the nature of that planet cannot be expected to influence the association.

A planet in natal Mercury aspect that has also a progressed aspect from Mercury will increase its strength potential to the power of both. A planet that is aspected in both charts by

Mercury of the other chart, either natally or by progression, increases its strength potential to the power of both.

As example, if in both charts Mercury aspects Venus, the two people will certainly have an emotional exchange, even without a Mercury-Venus progressed chart comparison aspect upon meeting. If in neither natal charts Mercury aspects Venus, but the other Mercury aspects are harmonious, a Mercury-Venus progressed aspect can show the development of love at that time. If neither natal charts have a Mercury aspect to Venus, and show conflict, a Mercury-Venus progression can show a period of love that can be strongly ambivalent, and of duration only during the period of the aspect.

If there is neither natal nor progressed Mercury aspect to Venus, we would not be justified in judging this to be an emotional relationship (even though people can relate very well and often marry without a Mercury-Venus in their compared charts). If the Mercury-Venus aspect is held by one person natally, and moves into progressed aspect by the other person, this denotes the time of the development of the emotional contact.

This illustration can be followed to the nature of each of the planets.

The orb limit is included in chapter 3, "Definition of Mercury Aspects". Progressed aspects are always 1 degree orb. Declinations are always 1 degree orb.

Just as in natal chart reading, the closer the aspect to its exact degree, the stronger the effect. By progression, the closer the aspect to its exact minute, the greater potential to action or event increases at that specific time.

Consider the times that Mercury changes station by direct or retrograde motion of longitude or declination for a change of condition, attitude, or event.

Consider the dignity or debility of Mercury and that of the planet aspected, for ease or difficulty of expression.

CHAPTER III

Definition of Aspect

THERE ARE TWELVE ASPECTS defined in the Mercury Method. They have basically the same meaning as in all chart reading, but specifically, they are geared to the particular affect they have between two people.

The person whose Mercury is applying to the planetary aspects will be named MERCURY. The person whose planet is receiving the aspect of Mercury will be referred to by the name of the planet.

Conjunction

When Mercury is in conjunction with a planet, MC or Asc, he has a familiarity with this person on a matter ruled by that planet. He understands or identifies from his own personal experience. They come together to share, or to experience at the same time, a quality or condition of that planet. The aspect neither attracts nor repels, it indicates neither endurance or separation. Mercury acknowledges the person on the level of the matters ruled by the planet.

The orb is 10 degrees, 12 degrees for Mercury to Sun, Moon and Mercury.

Opposition

When Mercury is in opposition to a planet, MC or Asc, he has something in common with this person on a matter ruled by that planet. Where they do reach an understanding or identification on this level, the full expression of the matter does not blend, so the relationship can only remain constant when held on the particular point in common. Where and when a larger adaptation is required they reach an irreconcilable impasse, from which they must withdraw until they can again contact on the common basis.

The opposition often indicates alternating periods of contact and separation. When the two choose a common environmental situation, a slow, inflexible struggle pulls them apart. There is at some time a separation of viewpoint or experience on the matters of the planet.

The orb is 10 degrees, 12 degrees for Mercury to Sun, Moon and Mercury.

Trine

When Mercury is trine a planet, MC or Asc, he spontaneously falls into contact with this person on a matter ruled by that planet. Circumstances and environment are such as to require little or no effort on Mercury's part. As a result, he is often unappreciative or indifferent. The benefit that Mercury accrues is commensurate with how positively he uses the capacity of the qualities of the planet involved. The Mercury who makes no effort may lose the benefit entirely.

The orb is 8 degrees, 10 degrees for Mercury to Sun, Moon and Mercury.

Sextile

When Mercury is sextile a planet, MC or Asc, he has the ability and the chance to express a matter ruled by the planet, with this person. There is a certain time when the two are drawn together easily and the opportunity is presented; if Mercury ignores this, the time passes and the chance is lost. The sextile in itself does not insure continuity or endurance, but does form an excellent basis for Mercury's freedom of choice in how to handle the matter.

The sextile is an aspect of more practical benefit and harmony than the trine. The trine conditions come easily whether Mercury has desire and interest or not, and he often will not make the effort to develop their full benefit. In the sextile, Mercury has the interest and is given the chance to pursue it.

The orb is 6 degrees, 8 degrees for Mercury to Sun, Moon and Mercury.

Semi-sextile

When Mercury is semi-sextile a planet, MC or Asc, he has a vivid spark of interest in this person on a matter ruled by the planet. This can develop only if both persons put an effort into cultivating its growth. Mercury is quite willing to make that effort, but can't carry it through without cooperation.

The orb is 2 degrees, 3 degrees for Mercury to Sun, Moon and Mercury.

Semi-square

When Mercury is semi-square a planet, MC or Asc, he has an intermittent contact with this person on a matter ruled by that planet. He does not fully understand that person's position or attitude. This causes Mercury a little worry and anxiety, that easily becomes criticism or irritation. The aspect brings slight changes that Mercury can handle constructively with the use of patience and firmness. The conditions right themselves in time by either losing importance or in being replaced by other circumstances.

The orb is 4 degrees, 5 degrees for Mercury to Sun, Moon and Mercury.

Inconjunct

When Mercury is inconjunct a planet, MC or Asc, he is strongly attracted to the person on a matter ruled by the planet. Without preamble, he takes an immediate interest and initiative developing the matter. In time, both his interest and initiative gradually disperse into other directions. Once stabilized, the contact can remain firm. By progression, the inconjunct indicates a period of climax and dispersement.

The orb is 2 degrees, 3 degrees for Mercury to Sun, Moon and Mercury.

Parallel

When Mercury is parallel a planet, MC or Asc, we find in all cases that this includes more than the interpersonal exchange between two people. It includes their present environment and their past experiences of conditioning. It includes the separate lives of the two people. It points out their similarities and their differences.

Though seen from Mercury's point of view, the action and the result of the parallel imply an exchange. The two people come together when they find something in common. They separate when their differences become too great. Because the parallel brings in their whole environment, in some cases it is the environment or other people who bring them together or who separate them.

The parallel is an intense and compelling contact that involves not only the outer life of the two people, but their covert and hidden areas. What appears on the surface and what's going on below the surface, may be two different matters.

The parallel indicates Mercury's contact with this person within the context of his individual life pattern.

North Parallel

When Mercury is N parallel a planet, MC or Asc, he has a different life pattern than this person on the matters ruled by the planet. Their previous environment and experience has been diverse, with some few similarities.

They come together with spontaneous circumstantial ease on a matter ruled by the planet. This matter hits a responsive cord in both persons. Mercury will develop this as long as it fits in with his primary life pattern. His majority of experience (of the matters of the planet's nature) is in an area apart from this person.

The N parallel does not indicate lack of affinity; often Mercury wishes passionately to interrelate on the matters of the planet, but circumstances are such that Mercury feels he cannot change the situation. The aspect therefore indicates a negative or passive contact, in that Mercury's efforts go into another direction rather than toward the person.

The N parallel is separative to the extent and in that area that Mercury must fulfill his primary life pattern apart from the person, on the matters ruled by the planet.

The N parallel by *progression* creates so impelling a bond that Mercury will change his patterns if possible in order to join the person on the matters of the planet. The slower moving planets in some cases hold the N parallel to the natal Mercury for many years. In these cases, we can find life-long associations that are highly satisfactory and productive in the matters of the planet.

Multiple N parallels create a bond that is strong enough to withstand the forces that tend to divide them.

South Parallel

When Mercury is S parallel a planet, MC or Asc, he has a different life pattern than this person on the matters ruled by the planet. Their previous environment and experience has had large areas of similarity.

They come together as a natural result of circumstances on a matter ruled by the planet. This matter implies a tension. Mercury has a diversity with the person but their bond is such that he is impelled to an expression of the planet's nature. Circumstances are such that Mercury feels he must make the necessary changes in order to work out an interaction with the person.

The S parallel is unifying to the extent and in that area that Mercury must fulfill a meaningful relationship with this person while maintaining his primary life pattern on the matters ruled by the planet. It is a positive, active contact, where the efforts of Mercury go toward the interaction.

Multiple S parallels create a tension of attraction that Mercury finds compelling.

Split Parallel

When Mercury is split parallel a planet, MC or Asc, his maximum contact with this person is on a matter ruled by the planet.

Each has a level of environment and experience apart from the other. They have areas that are uninteresting or incomprehensible to the other.

They can have a cohesive unity by maintaining the utmost candor and cooperation. They tend to be forced apart in direct ratio to any covert, coercive or hidden factors of inequality. These factors may come from one or the other, or from an outside force, or person, of that area in between them.

As an oversimplification, we may say that Mercury N parallel a planet wants to interact with the person on these matters but can't; Mercury S parallel a planet doesn't want to interact with the person on these matters but must.

With a N parallel, they have different backgrounds, with a S parallel, they have similar backgrounds.

With a split parallel, the greatest interaction between the two people is on the matter of the planet or at the *time* when they are both involved in the matters of the planet. Both persons have outside interests as well as their interaction.

CHAPTER IV

Mercury in Aspect to Mercury

THE FIRST CONTACT POINT between two people is their awareness of each other. The aspect of Mercury to Mercury is the primary indication of the ease, or the effort with which they meet. It shows the basic communication facility, the first level of expression. This is the only aspect that applies equally to both persons. Mercury in aspect to Mercury gives an innate, often below-conscious level understanding, a KNOWING, that is most marked in the conjunction and the parallels.

There are cases of relationship between people who have no natal Mercury to Mercury aspect. Until they have established experience in common and areas of interaction, their communication lacks the closeness of deep understanding. The first Mercury to Mercury aspect by progression that is made between the two charts tends to mark the nature of the communication until it is replaced by the next Mercury to Mercury aspect in their progressed chart comparison.

Mercury Conjunct Mercury

The two people have a familiarity. They can communicate on a level that is exclusive to themselves. They recognize similarities that they share, even to identical experiences with which they can identify.

In over half the examples, the two people had a mutual awareness and closeness in which they thought of each other as a "twin" or as an extension of themselves. In these cases, the two would spend many hours together, exchanging every thought and experience over periods of years. In other cases, where the age or interest gap was too large to build communication, there was still a recognition, or identification between them. An old woman with the Mercury conjunction Mercury of a youngster would continually say of him. "There's that nice Jones boy. That's my boy! Isn't he a fine lad."

In one case, of a woman (who also had Mercury split parallel Saturn and opposition Saturn) to a man, there was heavy antipathy. This was a business contact in which they had a close, tight communication of shared experience.

There are several romances in the examples. In these, the emphasis was on the exchange of understanding and KNOWING of each other.

The progressions of Mercury conjunct Mercury show more clearly a given period of close contact, or contact because of a similar situation. In one case, two women had the Mercury conjunction who were both lovers of the same man. They may have had a great deal of understanding in the identical experience, but there was not much affinity.

In the majority of the progressions, the two people would seek each other's company when possible. Each felt comfortable and that he was understood when with the other person.

Mercury Opposition Mercury

Each of the two people can express themselves with a level of understanding or identification with the other person. However, there is no blend of the same *type* of intellectual accumulations. They think about different things, and they think in different ways. The interests of each lead them into such diverse paths that neither can adapt to the environment of the other for longer than short periods. They can maintain association only under the conditions of alternate contact and separation.

The examples have such varied people as a traveling entertainer and his landlady, several cases of child-parent where the child left home early, business associates in different offices who occasionally worked together, a woman and her dentist. An actress with Mercury opposition the Mercury of her accountant said of him, "He points out all the things that I have never even thought of, yet I know exactly what he means."

There is one example of a long term marriage with Mercury opposition Mercury, 10 degrees. The woman had not the slightest idea of what her husband did each day. Neither could reason or explain the thought process of the other. Yet, they were both intelligent people, and they had a communication of understanding and identification for ten years.

The Mercury opposition Mercury progressions have more examples of intimate contact, as they include cases of friends and lovers.

They meet on a common ground, there is a vivid exchange of communication and understanding, then each returns to his own separate interests. The progression sometimes occurs to people who have a life long contact. It marks a time when their interests are divided to such an extent that there is a separation of viewpoint or experience, temporarily or permanently. In one such case, a woman left her husband for three months, to stay with their daughter who was having a baby. In other cases, there is frankly separation due to incompatibility.

Mercury Square Mercury

The two people are strongly drawn to each other. There is an initial obstacle, or incompatibility that stands between them. That difference may be in attitude or in circumstances. An exchange of understanding is needed for their breakthrough to communication. Once they make that breakthrough, a bond is built that is mutually stimulating. Until the two both make that effort, they will regard each other with a degree of wary formality.

These Mercury square Mercury examples include several of the finest relationships in the entire study, of people who have the deep close bond of enduring friendship, and of marriage. The bond includes problems and clashes, but the give-and-take implies a growth of understanding. These people overcame initial obstacles of some nature.

In one example, the very young mother of a teen-age daughter had Mercury square the Mercury of a man who knew them both. The two had a wary formality during the year they worked together in the same school. When the young man paid a visit to the home, an exchange was needed for their breakthrough to communication. As the woman was a direct person, she asked, "Are you here to visit my daughter or me?" The man said, "You." Two weeks later, they were married.

In many cases, the obstacle was age difference or a difference in station. In several cases, the two people were competitive in nature. The few who did not blend at all were people whose interests were too widely incompatible.

The Mercury square Mercury progressions put more emphasis on circumstantial obstacles. The two may be separated by distance to where they seldom see each other. One person may marry someone else, or be involved with someone else who is not agreeable to their friendship. In some examples, the two people have developed enmity from other situations but continue a cautiously formal contact.

Mercury Trine Mercury

The contact of the two people falls spontaneously into place. They have a natural and easy communication, as a result of similar interests. They have had some similarities in environment or experience.

The trine is an easy and usually pleasant contact, but a fairly indifferent one. The two do not have a depth of understanding, nor do they make an effort to develop it.

There are numerous Mercury trine Mercury examples of people who have long term associations in business, friendship, and marriage. The two people may grow up together or go to the same schools or go into the same business. They may have the same friends, or the same interests that bring them into contact continually.

In one case, a young woman who was accustomed to having lunch in a certain cafe, began to strike up an easy daily conversation with one of the waitresses. The two girls, who had Mercury trine Mercury, were three years apart in age, found that they lived within a few blocks of each other, and both liked to play cards and go to the movies. They easily formed a friendship and a social life in the same group of young people.

One girl later married the brother of the other girl and throughout their life they had a harmonious contact.

There are other examples of people who have a brief contact that comes easily, but they have no particular desire to continue. The trine insures the ease of circumstances and some interests in common. The other aspects have to be considered for desire and effort, and for depth.

The Mercury trine progressions are more placid than the natal examples. In many cases, the two people have no more than a casual contact that is pleasant. There are several brief love affairs that were mostly a case of the right circumstances at the right time. One had great intensity, as both persons also had Mercury to Venus aspects in a strong chart comparison.

Mercury Sextile Mercury

The two people are given a clear chance to develop communication and understanding. There is a time when they're drawn together easily. They have an interest in each other and in the same matters. With some measure of give-and-take, they can develop this into a close association. They must both make this effort, as the contact depends on a mutual interchange.

There is a wide range of effect in these examples. In all the Mercury sextile Mercury cases they come into the same environment with an interest in common. But due to other factors, one or the other may not choose to carry through with an effort, or response.

There are several close marriage and family relationships where the give-and-take was established to the benefit of both. There are also romances and friendships where the attraction and the interests in common were not strong enough to endure.

The Mercury sextile Mercury progressions also show some ease of opportunity balanced with some necessity of effort. In one example, two women met because their six year old sons played together. The two women found they had the same interests and enjoyed each other's company. As their husbands were not compatible to each other, they did not develop a blended social life, but the women often had morning coffee together. With this aspect, they began a friendship that lasted many years.

In some cases, the two people did not develop any understanding when they had the chance. They both went on to other interests and environments; when they met their contact was limited to, "How are you? Nice weather we're having." In other examples, both marriage and business contracts were formed on a mutual interchange of communication and effort.

Mercury Semi-sextile Mercury

The two people have a spark of interest in each other or in a common matter. If both make the effort to develop this, it can grow into an area of communication.

In the examples, where there is no other aspect but the semi-sextile, the two people may have nothing more in common than a similar fondness for artichokes. Where other aspects build substance and depth, the semi-sextile is a keen point in common that can grow into a bond.

We have this aspect between a mother and a young man who later became her son-in-law. Their initial interest in common was her daughter. This always remained their primary bond, and other aspects contributed to the other areas of their relationship. In the examples, the interest in common was a mutual friend, the same musical tastes, the same job or business, or the same attitude about some matter that affected them both.

In a like manner, the interest in each other is a spark that needs to be developed slowly and carefully into understanding.

Mercury Semi-square Mercury

The two people have an intermittent contact. They each have some area, or attitude, that the other has difficulty understanding, and here they have a break in communication. This area can cause some worry or anxiety, that may breed criticism or irritation. When they exchange a firm patience they can build a good communication on the matters of common interest.

In the examples, the two people learn to live with the worrisome area by ignoring it or occasionally by nagging on it.

There was one case of a father and a hippy son. Their contact was intermittent because of conflicting hours and interests. They certainly had difficulty from the time of the boy's childhood in understanding each other's attitudes in certain areas.

The boy lived at home, and he and his father shared an interest in antique automobiles. They worked together with patience and camaraderie on an old stripped down car. Then they would clash again, with the father saying, "When are you going to get a haircut? Why don't you wear shoes?", or the boy would say, "What do you mean, Nixon? Are you crazy or something?"

In all the examples the contact was intermittent, with friends, relatives, or business associates. In the progression examples, a marriage with Mercury semi-square Mercury showed a time when the two both had jobs with conflicting hours. Their understanding was hampered by short tempers and worry over finances.

A semi-square in some cases gives no more than a mild irritation over some attitude or a way of doing things. In other cases it shows a distinct worry over a specific misunderstanding.

Mercury Sesqui-square Mercury

The two people have a sudden situation, or intervals of disruption in their contact. Though this may agitate them both, it does break up the existing conditions to clear the way for a new line of effort. If they both decide to make that effort, they can build a relationship along different lines than before.

In the examples of Mercury sesqui-square Mercury where the two had a long term relationship, the aspect indicated occasional periods of disruption that put stress and changes to their contact. Periodically, a sudden situation caused agitation. These situations varied from job and time commitments that made a steady contact impossible; to being evicted from their house; to sudden attraction between persons married to someone else; to social or economic interference in a relationship.

By natal aspect or progression the sesqui-square indicates some change of condition at an indeterminate time, and in an unexpected manner.

Several people started romances with a sesqui-square, sometimes with an age difference or an interest gap that caused agitation. Other people separated with the sesqui-square and later reunited with more understanding. Others separated with relief.

Mercury Inconjunct Mercury

The two people have an initial awareness of each other that is similar to that of the conjunction. They can communicate on a level that is exclusive to them alone.

They do take an initial interest and initiative in developing their relationship but other matters gradually take the primary attention of both persons. Once stabilized, the contact can remain firm.

By progression, the inconjunct indicates a period of climax and dispersement.

The natal examples of the inconjunct were all of long term relationships that took no preamble, or build up. The people met with impact and settled down to life long associations in friendship, marriage and in family ties.

The progressions held less endurance. There were a number of short and snappy romances. There were social and business contacts of people who met with an instant facility of communication but soon drifted apart. There were two marriages that immediately settled into a routine, and one marriage that immediately disintegrated.

In the sesqui-square, the semi-sextile, semi-square, and inconjunct, other aspects must be weighed to get the complete picture.

Mercury Parallel Mercury

In all of the Mercury parallel Mercury examples, there is a recognition, a KNOWING, between the two people. This may never express in words, or even on a conscious level. But in the most casual to the most violent exchange, there is that awareness of the other person.

Out of 106 examples of Mercury parallel Mercury, we have only four in which we can find no apparent exchange of influence in each other's lives, in action or attitude. The examples range in importance from these four, to those of the most intense impact possible between two people.

Mercury North Parallel Mercury

The two people have different life patterns, with some few similarities. They are accustomed to diverse environments and associations. They meet by chance. There is an ease of circumstance, or a common interest that brings them into the same environment. Their initial acknowledgement of each other draws them together.

The two people can blend for a period of time, but the circumstances of their diversities are such that each must express his primary patterns in his own separate area. Only in the case of multiple N parallels of several planets is the bond strong enough to withstand the pressures that tend to divide them.

The examples include a warm friendship between a young Negro man and an older Caucasian woman, who met weekly for several years in a study class.

In one case, a man and woman from homes 3,000 miles apart met at a convention mid-way between. Their attraction was so intense, that they had a three day love affair, the impact of which stayed with them both for many years.

Another example was of two young men who were both musicians, who spent many compatible hours together smoking pot. One of them was world traveled and ambitious; he went on to active fields. The other man stayed in his peaceful world of marijuana and music.

In other cases, the bond of Mercury N parallel Mercury insured a life-long affinity, though the two people had such different lives and experiences that they rarely maintained enduring contact together.

The Mercury N parallel Mercury progressions include several of the closest friendships between people of diverse backgrounds, where the contact was broken by one person moving away. In some of the examples, there was a more casual and intermittent contact between people in widely varied fields of interest, who nonetheless made an impression upon each other's lives.

One progressed example was of a man and woman who fell in love. He was married, she was not. He was in a political office where divorce would jeopardize his public position. Their nationalities were different. At the time of this aspect, the man's Mercury made four N parallels to planets in the woman's chart; the woman's Mercury made two N parallels to the man's chart. In spite of all the pressures that would tend to separate them, the quantity of N parallels created too strong a bond.

He divorced his wife, resigned his public post to go into business, and they married. The marriage endured for their lifetimes.

In several other cases where the people were in love, their worlds were too different to include each other. They were unable to change their pattern structure, and they separated.

If both persons are able to change their personal situations to build common experience and communication, the exchange is beneficial to both.

In the majority of examples, the Mercury N parallel Mercury implies affinity, but without a depth of conscious understanding on a communication level. For the most part, it is a passive contact, where the efforts of both go into other directions rather than toward each other.

Mercury South Parallel Mercury

The two people have different life patterns, but with large areas of similarity that overlap. Their contact is the natural result of circumstances. It is an intense, meaningful exchange that affects them both, though in different ways.

All the examples but one were of long term relationships. The persons' lives overlapped in family ties, in lifelong friendships, marriages, in common interests that held them together. There were widely diverse basic patterns in either personality, career or experience. The two people are not at all alike, *except* in one outstanding area where they have something identical. They may have the same attitude and experience toward their parent or their child. They may share a passion for their common hobby or mundane ambition.

The matter in common is so important that it draws them together and holds them together, even when they feel that they have their separate lives to live.

By progression, the Mercury S parallel Mercury brings the two people together to a common experience that they both feel impelled toward. It is an intense contact that has a different result for each person.

The intensity of exchange is not always harmonious. The progressions hold one short-term example that made a life-long impression on both persons. This was the case of a man and a girl child. They came together by the natural circumstances of being neighbors. One night, when the children were all outside playing games in the summer darkness, the girl ran into the back porch of the man's home to play "hide". The man was drunk and he raped the girl.

The after effects were disastrous to both. There were tears, confusion, scandal, hysterics, police and three months of court procedure. The girl's family moved and the man spent a year in prison.

The identical experience that impelled these two together was the potential to violence that both had outstandingly indicated in their natal and progressed charts. It must be pointed out that there were many contributing aspects between these two. The man had 13 Mercury aspects to the girl's chart, including his Mercury one degree conjunction her Uranus. At the time of the attack, his Mercury was sextile her Mars, (action opportunity) split parallel her Mars, (maximum contact of trauma) and N parallel her Uranus (unconventional attraction).

The Mercury S parallel Mercury is not always discordant, but it does always imply some degree of tension and effort.

The two people may attempt to live together, or have an affair, or go into business together. In this, they have an impact on each other's lives that changes them both, but they cannot maintain a total integration. Both must fulfill separate life patterns of communication and experience while completing a meaningful exchange with each other.

Mercury Split Parallel Mercury

These two people have a maximum contact of identity with each other at some time, or over a period of time. Each has a level of environment and experience apart from the other. Each have areas that are uninteresting or incomprehensible to the other. They can develop the greatest unity when both have an open and sincere frankness and cooperation. They tend to be forced apart in direct ratio to any covert, coercive or hidden factors that enter the contact. These factors may come from one or the other, or from an outside force or person of that area in-between them.

In the examples we can find a unity between these two people that is so intense it approaches telepathy. They can speak each other's thoughts, feel each other's joy and pain, express each other's attitudes up to a certain point.

In the most casual or short term relationships, there is that time when the two people meet on a wide open level, with little of themselves withheld.

However, there is an area of their lives apart from each other. There is some attitude, perhaps deeply covert, or some experience that they cannot share with each other. The extent of this hidden area is the extent of the distance between them.

In many examples of the Mercury split parallel Mercury the area of personal experience that stands between the two is another person. In two cases, one of the persons was murdered. (These cases included the Mercury parallel Mars, Saturn and Neptune). In five cases, one of the two persons had a hidden lover or marriage that separated the two with the Mercury split parallel Mercury. The hidden areas also included certain attitudes (that could better be traced to the natal chart) that they felt they could not expose to each other.

When both of these people built a relationship upon candor and cooperation, there was an exceptional blending. In these examples, they both had to have room to express their identities in some area apart from each other, but they did not withhold communication or understanding.

The Mercury split parallel Mercury progressions indicate the time of maximum contact, when they blended in open cooperation, or were separated by a coercive force.

On page 28 is a Mercury chart comparison to illustrate the importance of the Mercury aspect to Mercury. The other aspects are given a cursory examination at this time.

The chart is that of two women, Joan and Pat.

Joan was a fashion model, a tall, striking blonde, with a slender face and body. She had a British accent from her childhood spent in England.

Pat was a secretary in a large shipping firm. She'd been raised in a small midwest town, and had come to the city several years before, to business school and job. She had short brown hair and a wide-eyed look of youthfulness.

(Mercury N parallel Mercury, progressed). From highly diverse backgrounds, the girls had come many miles to the same city. Their lives were widely dissimilar, both in work and social areas. Joan moved among a chic, modern crowd. Pat had been married to a young salesman for the past year.

They met one day while waiting in the dentist's office. There was an acknowledgment of each other that drew them together.

(Mercury semi-sextile Mercury, progressed). The two girls began a casual conversation. Pat was enchanted by Joan's accent and manner. Joan felt that Pat represented the round femininity that she lacked in appearance but identified with in nature. They made a date to have lunch together the next day.

(Mercury semi-square Mercury, natal). The girls had an intermittent contact over the next two years. The two N parallels drew them together in a responsive bond of acknowledgment and shared diversity. They certainly had areas of attitude that were less easy to understand, where there was a break in communication. Joan did not care for Pat's husband, and Pat did not approve of some of Joan's more sophisticated adventures.

The Importance of Mercury Aspect to Mercury

Natal	Joan to Pat	Natal	Pat to Joan
☿ 6 ♋ 22 18 N.20	∠ ☿ 1° ✶ Asc. 1° ☌ ♀ 4° ☍ ☽ 9°	☿ 21 ♌ 40 13 N.16	∠ ☿ 1° ☍ Asc. 3° △ ♅ 2° ⚺ ☽ 1°
Progressed 1960 ☿ 21 ♋ 15 ℞ 13 N.55 Natal ☿	N.♇ ☿ ⚺ ☿ -	Progressed 1960 3 ♎ 20 24 N.12 Natal ☿	N.♇ ♅ N.♇ ☿p ⚺ ☿p
Progressed 1962 ☿ 19 ♋ 45 12 N.10 Natal ☿	N.♇ ♀ □ ☿p	Progressed 1962 ☿ 6 ♎ 05 25 N.00 Natal	□ ☿ N.♇ ♅ -

(Mercury N parallel Uranus, progressed.) As this aspect moved toward its exact degree, the process of change was affecting both girls in their different life patterns. Joan went to work as a buyer for a national chain of stores, and she was considering the move to a better apartment. Pat was at combat point with her husband, and emotionally upset. She, too, had changed jobs, and went to work for a radio station.

(Mercury N parallel Venus, progressed.) In spite of their diverse social and emotional life patterns, Joan and Pat felt a responsive bond of affection that drew them together. (Mercury square Mercury, progressed.) They discussed sharing an apartment. There were obstacles for both of them. Joan would have to break a lease on her present place. Pat was still reluctant to break up her three years of marriage.

The strong bond of these three aspects impelled them together through the period of stress. They changed those parts of their life patterns that diversified, and moved into a new apartment.

(Mercury semi-sextile Moon, natal.) The two girls had enjoyed their periodic contact for some time, in which they'd shared many of the details of their everyday experiences and mental attitudes. (Mercury trine Uranus, natal.) Once the decision was made and the move established, Pat was excited about this new chapter in her life. She accepted it spontaneously, and easily moved into new areas of experience with Joan. Her position had taken Joan into a variety of contacts, and the two girls began moving into a social circle that was stimulating.

(Mercury conjunct Venus, natal.) Joan had a full understanding and identification with Pat's emotional and social life. She took pleasure in helping Pat to improve her style and make-up. The girls shopped together, sewed together, dated together, took drives together, shared their books and music and fun and games.

(Mercury sextile Asc, natal.) The shared affinity in a common environmental situation gave Joan the change of personal development as well as exchange.

(Mercury opposition Moon, Natal.) In the intimate details of their daily domestic life, Joan reached a penetrating understanding of Pat's mood and feeling. They had an active rapport of care and solicitude.

(Mercury N parallel Venus, progressed.) Joan did have areas of emotion that she had to express apart from Pat. Where the girls had affection for each other, they found that they did not like the same men. They had a clash in attitude and opinion whenever either of them moved out of the general social group and began dating one man. Before long, Joan discreetly began a steady relationship with a married man.

(Mercury opposition Asc, natal.) Though Pat had tastes, disposition and viewpoint in common with Joan, their basic personality and attitudes did not blend in full.

(Mercury N parallel Uranus, progressed.) Pat began to express an independence that she'd not had the courage or inner stability to develop before. The clerical work that she was trained for palled before her growing interest in the technical fields that she saw around her at the radio station. Pat applied for an assistant position in the recording department, and began an active study in communications.

With the two natal oppositions and the separate areas of emotional expression for Joan, and expression of original change for Pat, the two girls began pulling apart.

After ten months together, Pat moved out. Their break was not antipathetic, as they both appreciated the closeness they had felt for each other. There simply was too much diversity in basic attitudes, personality, structural and emotional background for a total blend in a common environment.

CHAPTER V

Mercury in Aspect to Sun

THE SUN RULES the ego significance and the integral character of individuality. It is the drive of importance, authority and worldly goals.

Mercury makes contact with the Sun on these matters. He relates to the Sun. He connects in thought or meaning to the goals and authority of this person. The pride, the self esteem and conscienciousness of both are involved in a mature and vital relationship.

When Mercury and the Sun clash, it indicates a competitive conflict of ego and power drives. They both need to develop respect and understanding.

The Mercury-Sun aspects frequently pertain to business relationships, and are often present at the contracting of marriage or partnership. It connotes a contact on matters of importance to both.

The Mercury-Sun parallel is a compelling and significant relationship that brings into sharp contrast the similarities and differences of goal in the life patterns of the two people.

Mercury Conjunct Sun

Mercury has a familiarity with the ego drive of the Sun. He relates to the Sun as an equal, giving deference and respect to the authority and importance of the Sun. They will share, or experience at the same time a common goal, or a relationship that is important to both.

Mercury can understand or identify with the vital goals of the Sun from his own personal experience. This is an impartial acknowledgment, as the conjunction is neutral.

In the examples, even when Mercury has the greater role of authority, he pays attention to the importance of the Sun, and gives him credit for his goals or achievements. In all cases, Mercury gives some support or deference to the prestige or position of the Sun. He does so quite naturally, even casually. They have some mundane matter in common, or have had experience in the same business.

In one example, of a woman with Mercury conjunct a man's Sun, they met when she was teaching a class that he subsequently taught. They had a common interest in the same subject matter, as well as a simultaneous experience in that both were having marital problems.

In the case of two women with Mercury conjunct Sun, Mercury said, "Yes, I know when she comes in; I can feel her authority. I don't understand her emotional hang-ups, but I understand what she wants to do."

In all cases, the two have either a relationship with the same person, or a similarity of relationship at the same time. This is not an identical situation, but one that is akin.

When both person's Mercury is conjunct the Sun of the other, it is a vital relationship that is meaningful to both. The other examples are more casual.

The progressions indicate a time when Mercury identifies with the significance of the Sun, on a business matter or a matter having to do with the worldly position or prestige. Mercury is proud of the Sun and feels that his own importance is enhanced by being with the Sun.

Mercury Opposition Sun

Mercury reaches an understanding or identification with the Sun for certain goals and interests held in common. They share some degree of integral character that is meaningful to both. They share some matter of worldly goal or position in common.

However, the full expression of that goal, the manner of achievement, or the deep ego drives, do not blend in full. Mercury cannot fully adapt to the matters that are important to the Sun. When the time comes for a larger adaptation, they reach an impasse, from which Mercury must withdraw until he can again make contact on a common basis.

The examples contain family relationships where the Mercury child eventually drew away from the Sun parent to find his own individuality and meanings. In the business relationships, Mercury related most agreeably to the Sun in extending a maximum of coopèration and trust.

The contact remains the most consistent in quality where it is intermittent in time. When the two choose a common environment, a slow inflexible struggle of character and meanings pushes them apart.

Outside of the child-parent relationships which were for the most part agreeable, the longest example of the Mercury opposition Sun was a 12 year marriage. The last five years of that marriage was a gradual pulling apart, as the interests and ego drives of the two became as oil and water.

The Mercury opposition Sun progressions show the period of full contact on a matter of significance either in goals, or relationship, with a subsequent separation. While this was final only in a few cases, it often showed that circumstances would tend to divide the two.

In an example, a woman went to work for a man with whom she had Mercury opposition Sun. Where they had understanding and identification in their common interest in the office, Mercury felt that the Sun's business methods were lacking integrity, so she quit.

In a personal relationship of a man with Mercury opposition a woman's Sun, they had a romance-friendship. They both were ambitious in their personal goals, they had intellectual interests in common, and had character similarities. Mercury would make an intense contact with the Sun, but they could not fully blend. An ego conflict would develop, a discrepancy in meanings would become irreconcilable. Mercury would withdraw for weeks or months before again calling the Sun.

There is only one consummate romance in the progressions. As the aspect moved out, the two married.

Mercury Square Sun

Mercury is strongly drawn to the authority and strength of character of the Sun, but there is an ego incompatibility that implies conflict. Each has importance and significance in their own environment; when together they tend to clash.

When Mercury gives a deferential respect to the capacity and interests of the Sun, he receives the Sun's magnanimity.

When Mercury takes wilfull leadership, he either has a power struggle or loss of contact with the Sun.

These examples include both long and short term relationships, most of which were deep and meaningful contacts. Where Mercury is a big enough person to allow the Sun to express his ego, authority and capacities, it is a rich association. But often Mercury is stimulated to ego-competetiveness with the Sun that starts a period of push and pull conflict of domination.

When people meet during a Mercury square Sun progression, Mercury is quite willing to break through obstacles to relate to the Sun. If the progression occurs in an established relationship, it indicates a period of conflict or hostility.

There are more personal than business relationships in these examples, though the worldly position or business touches upon the contact. Many of the natal and progressed examples of Mercury square Sun are of people who meet through work or business. There are several doctor-patient, teacher-student relationships; there are others where Mercury or the Sun sells a service or product.

In all cases, Mercury does relate to the Sun. The maturity and significance of the relationship depends for the most part on his attitude.

Mercury Trine Sun

Mercury spontaneously relates to the Sun, as environment provides the natural opportunity and impulse. He easily gives deference and credit for both the character and position of the Sun. As the relationship comes so easily and naturally, Mercury is not always appreciative. When he makes no effort to develop a deeper understanding, or when he has an ego or integrity clash with the Sun, the contact dissolves as easily as it formed.

The examples show many relationships for common goals in business, marriage, and friendship. Most of them endure pleasantly and beneficially. There is an easy and natural affinity, as Mercury is proud of the Sun.

In those cases where there is an ego or integrity clash, or where stress comes from other aspects, they both tend to move apart from each other rather than hold a steady unity. When the events do not fall naturally into place, Mercury does not seem to care enough, or know how to make the appropriate effort at the opportune time. As a result, the progressions in half the cases are as indicative of separation as of blending. Where the Mercury trine Sun is exceptionally harmonious, it does seem to lack depth.

Mercury Sextile Sun

Mercury can understand and appreciate the goals and character of the Sun. Circumstances give Mercury the opportunity to develop a relationship based on an exchange of respect and credit. They are drawn together easily at a time when they can mutually benefit each other. If Mercury does not take the initiative, the time passes and the matter loses importance.

All of these examples were harmonious. Mercury had an interest and respect for the Sun, and there were certain goals and interests in common. The aspects pertain largely to business relationships, or where the two people are involved in a goal that was prestigious or to their credit.

As example, a man's Mercury made the sextile to a young woman's Sun. In spite of her youth and business inexperience, he offered her the management of his restaurant, which she handled well to their mutual credit.

In an example where Mercury did not take the initiative, a woman had Mercury sextile the Sun of her employer. As she was reticent to assert her capacities, she stepped back from the chance to take a more authoritive position and the employer gave the job to someone else.

In all cases, there was a distinct period of time when Mercury has the chance to relate to the Sun. It is more indicative of business and goals than of personal relatedness.

Mercury Semi-sextile Sun

Mercury has a keen spark of interest in the character of the Sun. As they have some interest in common, with mutual effort they can develop a relationship.

The examples of Mercury semi-sextile Sun must be considered with the other aspects.

In one case, a woman's Mercury was semi-sextile a man's Sun, where she applied for a job in his company. She had a vivid spark of interest in both the job and in him personally. He only had one Mercury aspect to her chart, so couldn't care enough to make either effort or response. Nothing came of it.

In another example, two men who were life long friends had the Mercury semi-sextile Sun. The other aspects that contributed assured their contact and the man with Mercury often deferred to his friend's judgment and decision in matters of common interest.

The progressions referred to specific times when that spark of interest either began a contact, or stimulated slight changes in the relationship.

Mercury Semi-square Sun

Mercury has an intermittent contact with the Sun, in which he relates to the Sun's significance. As Mercury does not fully understand the Sun's position, the relationship may vacillate. Mercury may have some worry or anxiety about a certain matter or personal attitude of the Sun. The matter will either right itself or lose importance in time, so Mercury can be patient and firm.

The examples of long term relationship indicate periods of intermittent concern. A father with Mercury semi-square his daughter's Sun had periods of worry about her health during her childhood. He was concerned on occasion about her grades and deportment during school years, and was anxious about her romances during her young adult years. Each time the matters straightened out with time, and Mercury's firm patience carried him through.

In another case of a long term marriage, the wife's Mercury was semi-square her husband's Sun. She occasionally had concern over his extra marital affairs, but he always came home.

The progressions show a given period of time when Mercury is not sure of the Sun's attitude toward him, or does not know what's happening with the Sun's business or personal affairs.

There were several brief romances where Mercury had no idea of the motive or intent of the Sun. Mercury worried about this, and in time he learned there were other factors involved.

Mercury Sesqui-square Sun

Mercury has a sudden situation, or intervals of disruption with the Sun, that break up the existing conditions of relationships. This causes some degree of agitation to both. Mercury needs to make a decision of whether to begin, or continue a relationship with the Sun. With a little effort, he can connect with understanding to the goals and significance of the Sun on a different level than formerly.

In many of these examples, the disruptive situation involved a third party. Mercury separated from her husband, and reformed a relationship on a different basis with her friend, the Sun.

Mercury's brother was killed, in a case where he had a sesqui-square to the Sun of his brother's widow. He related with support and deference to her changing goals and meanings. It was a period of agitation for them both.

Mercury fell in love with a girl whom his boss was dating. With the sesqui-square to her Sun, this was agitating to them both. With a little effort, he reached her understanding and they began a relationship.

An actress with Mercury sesqui-square her male friend married several other men over a period of years. The friendship had periods of disruption, but it endured.

In other cases, the disruption was between the two people over intervals of changing goals and meanings that required readjustment in their relationships.

The progressions often showed worldly contacts at a time of business disruption that was agitating.

Several personal relationships were begun, with some indecision on Mercury's part. In one case, a man with Mercury sesqui-square his wife's Sun had an affair with another, causing a period of agitation. He did relate to his wife with pride and esteem, and they reformed their relationship on a deeper basis than before.

The Mercury sesqui-square Sun aspect does involve the integral character of both, but it tends more to personal relationships than business. In the business contacts, the two people usually separated over this disturbance.

Mercury Inconjunct Sun

Mercury relates strongly to the goals and authority of the Sun. He takes initiative in developing the contact, as they have matters of common significance. In time, Mercury's interest disperses into other areas. Once stabilized, it can remain a firm relationship.

The progression indicates a period of climax and subsequent dispersement.

In the natal examples, Mercury related to the Sun with no question. He accepted the authority and importance of the Sun from the onset. In time, his attention went to matters of other significance, but he retained a respectful relationship to the Sun that in most cases was long term.

The progressions show a period of importance to Mercury. He may relate to the extent of his capacity, but it's either too late or not enough. In only one example was a relationship begun. Several did endure through a time of changing goals, but the majority reached a climax that was separative. Several marriages were in their final period of struggle; in most of the cases, the Sun withdrew.

Mercury North Parallel Sun

The life pattern of Mercury's worldly goals differ from those of the Sun. Their environment and experiences in position, in importance and in meanings have been diverse.

They come together with spontaneous circumstantial ease as Mercury recognizes in the Sun a quality or condition to which he relates. There is a responsive exchange. They can develop a bond in ratio to their similarity in integral character.

However, the circumstances of their diversities are such that Mercury is impelled to express his goals and meanings in his now familiar area.

The examples of Mercury N parallel Sun are of relationships vital to the character and meanings of both persons.

In one case, a woman who was oriented to cultural and artistic values had Mercury N parallel Sun to a man who was steeped in political environment. With a deep integral bond, they married. While the man achieved political prominence, his wife kept pace with his prestige, but in her own pattern of cultural development. After his death, she maintained his image in relatedness, but continued her pattern that was quite diverse from his.

In other examples, Mercury related in deep bonds of friendship during periods of similar experience, but eventually went on to his own goals and life patterns after a year or two at most.

In several cases, Mercury found the integrity patterns of the Sun to be totally incommensurate with his own. In these cases, the Mercury N parallel Sun was highly antipathetic.

In the progressions Mercury approached the Sun easily on a certain goal or interest they had in common. He developed this relationship as long as the response fit his primary life pattern. His maximum worldly goals and integral relationships were expressed in an area apart from the Sun.

The business contacts were most productive, as Mercury could maintain this bond while living his separate life. There are examples of both an actor to his agent and an agent to his client during a prestigious period. There are business partners whose integrity in common was their sharing of illegal practices.

When a child has the aspect to a parent, it indicates a period of growth for the child's interests and goals to be directed into other areas than those of the parental values.

The relationships of Mercury N parallel Sun are separative only to the extent that Mercury must express his primary drives of importance and significance in other areas. There is the bond of a matter in common.

In some few cases, where there are multiple N parallels, Mercury will alter his life patterns in order to remain with the Sun.

Mercury South Parallel Sun

Mercury has a different life pattern of worldly goals and of integral character than that of the Sun, but with large areas of similarity that overlap.

They come together as a natural result of circumstances and environment. They have a great deal in common, in their drives of importance, their worldly goals, or their integral character of individuality. These matters affect them both in vital areas of their life's experience, though in different ways.

The Mercury S parallel Sun progressions indicate a tension that impells Mercury to relate to the Sun.

In the examples, the two people may be in the same business with a common drive to achievement that they can share and understand in each other, while having quite diverse personal lives. Alternately, they may have highly diverse worldly goals in the business sense, but a deep bond based on similar integrity patterns, or relationships in common.

In some cases, the bond is familiar, where Mercury is impelled by common heritage to relate to the Sun.

In a progressed example of a man and woman who were divorced, the man's Mercury was S parallel his former wife's Sun. He was impelled to relate to her authority and goals as they both had a love for their children, which they expressed in different ways.

In another case, a woman's Mercury was S parallel her husband's Sun at the time of his death. The tension of this vital matter impelled the woman to expend all of her energies in relating to his mundane affairs.

There was only one example of a romance in the Mercury S parallel Sun charts, in which a young woman related to the authority and dominion of an older man with whom she worked. The affair began and ended during the four years that the progression was in effect.

In several examples, the business matter in common involved large amounts of money and prestige. One case was of a man who is world known for his achievements, with Mercury S parallel the Sun of one of his co-workers. This man also had a great deal of authority in his position. They became business partners at the time of this progression, with a half million dollar investment.

As in the majority of Mercury aspects to the Sun, the S parallel is most favorable to business contacts or to those that connote a meaning for goal.

Mercury can fulfill an important relationship with the Sun while maintaining his own integral individuality.

Mercury Split Parallel Sun

Mercury relates to the Sun with the maximum of his ego-drive of individuality. At some time, or over a period of time, Mercury's experience with the Sun is one of the vital relationships of his life.

However, they both have a level of goals and importance in areas apart from each other. These areas are uninteresting or incomprehensible to the other person.

Where they are impelled together, Mercury will remain with the Sun only to the extent that their integral character is based on the same goals and meanings. When their separate areas of interest hold incommensurate factors, they tend to be forced apart.

The examples are of intense contact, both in business and personal matters, but they do not tend often to relationships of solid duration in a common environment.

A man with Mercury split parallel the Sun of his brother related to him intensely. They were impelled together periodically, but then returned to their own environment.

A woman with Mercury split parallel the Sun of a man in the same business began a romance with him. Though she was strongly attracted to him, she could not consummate an affair, as they both would veer off to other interests and relationships whenever they reached a maximum contact.

The business relationships fared better, as they gave prestige and credit to both persons. The intermittent contact maintained the most unity.

The progressions clearly indicate the period of time when Mercury reaches a maximum relationship to the Sun. They are impelled together by a business or personal goal. But where there is a bond, there is also a prohibition. Where there is a similarity in character, there is an ego-conflict. Where there is a similar goal, there is diverse experience. In very few cases are the two able to completely unify.

In the example of two young people in love, the boy's Mercury was split parallel the girl's Sun. They related to each other with the utmost candor and trust, but his school and family duties held them apart much of the time.

In other cases, the two people met with impact, but neither could adapt to the environment of the other.

Many times, the relationship to a third person stood between them.

In the long term relationships, the progressed Mercury split parallel Sun indicated a time when both persons developed goals and experiences apart from the other, that were difficult to share.

The Mercury Chart Comparison on page 43 illustrates a business exchange. Buddy is an actor who went to work at the age of five years. The agent's name was Reba.

Buddy was a self assured boy with the type of face that is usually called a mug. He had a wide mouth, round blue eyes and a bridge of freckles across his pug nose; a stocky, firm stance and a natural appeal that came across on camera.

The Mercury Chart Comparison in a Business Exchange

	Reba to Buddy		Buddy to Reba
Natal ☿ 13 ♈ 34 4 N.15	□ ☿ 7° ⚹ ☉ 5° ⚹ ♃ 6° ☍ ♄ 7°	Natal ☿ 6 ♋ 18 25 N.21	□ ☿ 7° □ ☉ 2° △ ♃ 1° ⚹ ♄ 6° ⚺ ♇ 1° ⚺ ♆ 1°
Progressed ☿ 17 ♉ 59 14 N.04	⚹ ☿ ⚺ ☉ ⚻ ♆ ☌ MC N. ♇ ♃	Progressed ☿ 17 ♋ 24 22 N.10	⚹ ☿ □ ♃

Reba had been in the theatrical business for 20 years as an actress and an agent. When Buddy came into her office with his mother, she signed him to an initial contract without hesitation.

Buddy did his first job within a month; a 60 second commercial for a national product. He did not achieve overwhelming fame or fortune, but was a good working child who built a log of credits in TV drama and commercials.

(Mercury square Mercury, natal.) The initial obstacle between Buddy and Reba was their difference in age and station. When Reba (with an Aries Sun sign) made a direct and open appeal to Buddy's (Leo Sun sign) sense of drama, their breakthrough to communication was established.

(Mercury sextile Sun, natal.) Reba related to Buddy with confidence and assurance of his capabilities.

(Mercury sextile Jupiter, natal.) She gave Buddy her time, her interest and enthusiasm, her loyalty, and the benefit of her experience in representing talent. The contact expanded Buddy's experience, and was a financial benefit to them both.

(Mercury opposition Saturn, natal.) This was a business relationship. Reba was a hard working, responsible person. The possibility of financial loss with Buddy was alleviated by the Jupiter aspect, but the Mercury opposition Saturn expressed after two years in that Reba lost Buddy's contract renewal. Their separation at that time was a grief to them both.

(Mercury square Sun, natal.) Where Buddy was highly admiring of Reba, his dominant Leo nature did clash with her decisive Aries. Each had importance in their own environment. When Buddy took the center of the stage in Reba's office, she would firmly put him in his place.

(Mercury trine Jupiter, natal.) Buddy was not old enough to fully express the benevolence and loyalty of this aspect. He did feel the burst of good will and pleasure in Reba's company, to where he would greet her happily and devotedly.

Financially, it was a highly beneficial exchange, as his two years work gave Buddy a savings account of well over $20,000. Reba collected 10% of this.

(Mercury sextile Saturn, natal.) Buddy was given the opportunity to work with Reba, and for all of his five years of age, he did a responsible and capable job. The stronger aspect of Jupiter (one degree) overweighed the problematical possibility.

(Mercury semi-sextile Neptune, natal.) Buddy was fascinated by the theatrical world that Reba represented. They both put the effort into developing a contact on this matter.

(Mercury semi-sextile Pluto, natal.) The only aspect that would inhibit Buddy's cooperation with Reba was the infrequent clash of Mercury square Sun. He and Reba were often in groups working together for a common goal.

(Mercury sextile Mercury, progressed.) When Buddy and Reba met, they were drawn together easily with an interest in each other and the same matters. The natal Jupiter that they both had assured an exchange of good will, to where they both made the effort to build communication and understanding.

(Mercury semi-sextile Sun, progressed.) This spark added weight to the natal Sun aspect they both had. Reba related to Buddy with complete assurance.

(Mercury N parallel Jupiter, progressed.) Reba's business and financial affairs had been, were, and continued to be in a larger area of interest apart from Buddy. They hit a mutual responsive cord of affinity and exchange of benefit. Buddy's residual income continued to accumulate after the two years of their business association.

(Mercury inconjunct Neptune, progressed.) Reba had an instant acknowledgment of Buddy's capacity to perform before the camera.

(Mercury conjunct MC, progressed.) Reba could well identify with Buddy's career and reputation. She acknowledged and developed his public prestige as part of her job in representing him. They appeared together in public situations, and in press releases.

(Mercury square Jupiter, progressed.) Buddy adored Reba. When the new experiences of stage, crew, lights, camera and direction threatened to be just too much for him to handle, Reba would boost his ego, praise his capacity and fire his enthusiasm.

Buddy and Reba separated after two years together. Buddy's Mercury moved into a split parallel to Reba's Uranus. Changing events affected them both.

Buddy's mother was remarried at this time, to a man whose work transferred them to the opposite coast. Buddy was losing his teeth, and gaining height. His jobs had diminished for several months. His mother felt that his career was less important than the home she chose to make with a man whom Buddy liked.

Reba was enlarging her office at a new location.

Had their patterns of change been together, it would have tended to unite them. As it was, Buddy and Reba moved into a new chapter of life patterns apart from each other. The change itself was the climax, or maximum of their relationship at that time.

CHAPTER VI

Mercury in Aspect to Moon

THE MOON RULES the domestic life and the fluctuating details of everyday experience. It is the subliminal, instinctive mental process that determines mood and feeling.

Mercury makes contact with the Moon on these matters. He often has a penetrating understanding of the unconscious mental processes of the Moon, that implies a responsive exchange of adaptation.

Though the contact tends to fluctuate, it is marked by meter. There may be daily contact, or periodic contact in various patterns.

The largest percentage of Mercury-Moon aspects indicate domestic relationships or contact in a home environment. It is markedly harmonious, implying for the most part affinity in all aspects but the square, where there is some measure of conflicting time or mood.

As the Moon rules the functional body systems, the health or welfare of one of the persons may require attention. In many of the aspects there is an exchange of nourishment or care, tenderness or protection.

The Mercury-Moon Parallels indicate the detailed or instinctive relationship of the two people, within the larger context of their separate life pattern of domesticity.

Mercury Conjunct Moon

Mercury has an instinctive familiarity with the moods and feelings of the Moon. He understands or identifies with the Moon's mental processes.

Whether Mercury shares with the Moon a long period of close daily association, or intermittent periods of contact, he is inclined to responsive sympatico.

They will share, or experience at the same time a domestic relationship or situation. Mercury has at some time, or often, a similar domestic situation to that of the Moon.

The majority of examples were family or family friends, in-laws, or friends living together.

The conjunction implies a certain objectivity to all the Mercury conjunction of planets except that of Mercury conjunct Moon. In these examples, Mercury feels and responds in feeling to the Moon. There is usually an exchange of protection and trust, nourishment and care of each other.

A five year old boy with Mercury conjunct his mother's Moon expressed this solicitude and understanding when he said, "Mom, you'd better go to bed early tonight so you're not in a bad mood tomorrow."

Two adult women with Mercury conjunct Moon shared many years of domestic crisis, joys and sorrows. They would both pick-up and respond to each other's mood and mental attitudes.

In one example, Mercury, an actress, married the man who had previously been married to the Moon. The two women experienced in this way a similar domestic relationship and situation. There was a periodic contact, as both women bore the man children.

The progressions show a period of time when there is periodic contact, either on a daily or intermittent basis. In all of the examples, there is more or less of the sharing of the intimate details of the mental processes and the day to day experiences.

A woman had Mercury conjunct the Moon of her husband's best friend. They had a warm and easy exchange of sympatico, in that they both understood quite well her husband's bad and good qualities.

There are very few marriages in the examples. The male-female contacts tend more to a fraternal relationship.

Mercury Opposition Moon

Mercury has certain mental processes in common with those of the Moon, to where he has a penetrating understanding of the Moon's mood or feeling.

However, their mental attitudes do not blend in full. When they reach an understanding or identification in the exchange of the intimate details of everyday experience, they reach an impasse from which Mercury must withdraw.

They come together in a domestic or periodic contact. Mercury has a solicitude and concern for the health or welfare of the Moon. As he cannot respond in full adaptation to the moods and feelings of the Moon, there is a periodic separation.

The contact can endure on a fluctuating basis. In a common environment, the two become irreconcilable to where they are separated by mood and circumstances.

The examples retain the most harmony in the familial relationships where the two do not live together, as with in-laws or a parent living apart from the child.

The several marriages and romances in these examples become irreconcilable. In each case, Mercury withdrew from the fluctuating moods and overemotional feelings of the Moon, either periodically or permanently.

In several examples of Mercury opposition Moon, the two people had a fraternal love for each other, where they shared mental interests over periods of years. They spent many hours in each other's homes in an exchange of solicitude, nourishment and understanding.

There was one example of a dentist with Mercury opposition the Moon of his patient. He had a concern for her welfare in their periods of contact.

In another example, two actors with Mercury opposition Moon played in a "Ben Casey" TV episode where Mercury took the part of the doctor caring for the Moon as his patient.

Mercury Square the Moon

Mercury is strongly drawn to respond to the mood or feeling of the Moon, but there is an incompatibility in their instinctive mental processes. Where Mercury does have concern for the Moon's physical or mental health or welfare, there are obstacles to block his sympatico.

When Mercury is gentle and protective of the Moon, there is a responsive exchange.

When Mercury is temperamental and unsympathetic, the periodic contact will fluctuate or break entirely.

There are only four examples where the contact is not set in the home environment.

One of these was a brief affair in Mercury's hotel room; two are doctor-patient relationships, where the doctor also treated Mercury's children; and one was of a young man and older woman who met in the school class room, had a brief exchange of sympatico, and separated.

The majority are family relationships, neighbors, people who lived together, or people who are brought together by a member of the household.

In one example, a man's Mercury was square the Moon of both his sisters. He had a familial response of loyalty to both, but his mental processes and attitudes differed from theirs. Even as children, their everyday experiences took different patterns; their moods and temperament conflicted. As adults, Mercury had periods of deep concern over the mental and physical health of one sister. During their life time, their periodic contact vacillated, both in timing and in sympatico.

There are beautiful examples of firm relationships where there is an equal exchange of nourishment and care, trust and tenderness.

The Mercury square Moon progressions show a period of time when Mercury is drawn into a domestic contact with the Moon, or has concern for his health and welfare. In certain examples, the Moon was pregnant; was homeless; had a physical or mental health problem. In other examples, Mercury was drawn into the home environment of the Moon where there was some obstacle to overcome in time or distance, or in diversity of mood.

Mercury Trine Moon

Mercury has a natural, easy and spontaneous response to the mood and feeling of the Moon. Though he has no depth of understanding of the mental processes of the Moon, he can adapt instinctively, with an offer of protection and trust, solicitude and tenderness.

Mercury accepts without question a closely detailed, or periodic, contact with the Moon that is often domestic in nature or set in a home environment.

At some time, or periodically, the everyday experiences of the two diversify, so the contact fluctuates in mood and timing.

Some of these examples are of people with intensity of contact that can be traced to other aspects. The Mercury trine Moon aspect by itself shows a casual acceptance of benefit.

Where Mercury adapts to the Moon with appreciation, he receives in turn the greatest nourishment and care. Where Mercury makes no effort of sympatico, the contact is inconsistant or simply dissolves.

In one example, a Japanese student had Mercury trine the Moon of the American woman who took him into her home as a member of the family. Though there was a language barrier, Mercury was a Pisces, and he instinctively responded with consideration and appreciation.

The two people developed an exchange of tenderness and care, protection and trust that was mutually nourishing for the three years that they lived together.

In another example, a mother had Mercury trine her daughter's Moon. Though she did not understand her daughter's mental processes, she made the instinctive maternal effort to respond to her daughter's mood and feelings with trust and solicitude. In this way, they developed an easy harmony.

One of the examples of close daily contact was set in the home of the Moon, the White House, Washington, D.C. The relationship was not familial; it had to do with the domestic affairs of a nation.

In one case, a man's Mercury was trine the Moon of his girl friend. During their six month affair, he accepted easily the intimacy of mood and feeling without making any effort to deeper understanding or appreciation. When she became pregnant, Mercury accepted this with a casual solicitude and helped arrange an abortion. The mood of the relationship changed, and they separated.

Three of the examples are of an adopted child to a parent, that gave the child a spontaneous domestic benefit.

Three of the examples are of doctor-patient relationships, where the periodic contact involved the benefit of care and protection.

The progressions are more casual, indicating a period of time when Mercury can easily have a periodic, or domestic contact with the Moon with little or no effort.

Mercury Sextile Moon

Mercury has the ability to respond to the mood and feeling of the Moon. There is a certain time when the two are drawn together easily and the chance is clearly presented for Mercury to offer solicitude and trust to the Moon.

When Mercury accepts this opportunity, the two develop a greater or lesser degree of exchange of the everyday details of experience, with mutual sympatico and appreciation. When Mercury does not make this effort, the contact gradually loses affinity.

In some of the examples of friends and of relatives, there was an intimate exchange of penetrating understanding. In other examples where Mercury made no effort to adapt, he quite lost any chance of affinity with the Moon.

Two women lived as neighbors for ten years with Mercury sextile Moon. Mercury was accustomed to a quiet, genteel life. When the Moon moved into her home, she brought with her music and children, noise and activity. Mercury chose to snub the Moon, so whatever sympatico they could have had was stillborn.

In several examples, the two people developed intense romances, with a penetrating exchange of adaptation. A woman with Mercury sextile her husband's Moon said, "I know so well how he feels, and he takes care of me as no one ever has in my life. I'm closer to him than anyone in this world."

There are child-parent examples and doctor-patient examples.

The progressions show the period of time when the opportunity is spontaneous. The choice and effort is up to Mercury.

Mercury Semi-sextile Moon

Mercury has a vivid spark of response and sympatico with the Moon. If both persons put an effort into the development of trust, they can build a harmonious periodic contact.

The majority of these examples are harmonious. The periodic contact ranged from daily over a period of years with family, neighbors and friends, to intermittent contact over the span of a lifetime.

The other aspects must be considered to fill out the picture. The Mercury semi-sextile Moon by itself alone would give no more than an occasional responsive contact.

Mercury Semi-square Moon

Mercury has an intermittent contact with the Moon that vacillates a great deal in time and tempo. Where Mercury responds to the mood and feelings of the Moon, he does not fully understand the Moon's mental processes. This causes Mercury worry or anxiety that exasperates his patience. With a little effort to firmness, Mercury can handle the fluctuation in their contact.

The examples were mainly of long term contacts, usually in the home, occasionally in the office.

In one example, a woman's Mercury was semi-square the Moon of her daughter-in-law. As the young people lived some distance away, they usually saw each other during holidays and visiting occasions. Mercury liked her daughter-in-law, but she had to contain an impatience at the younger woman's manner of housekeeping. Mercury also worried about their diet, health and welfare.

In another example, a woman with Mercury semi-square her husband's Moon worried about his health and wellbeing while he was away; she simply could not understand why he felt he had to join the Army.

In the progressions, the fluctuation of contact was more disturbing to Mercury, who in several cases felt that he was losing contact with the Moon. In one example, the two people separated for a period of time.

In another example, the two people met, Mercury made an attempt to relate to the Moon, but there was not enough understanding to offset the vacillation in mood and feeling and they separated.

Mercury Sesqui-square Moon

Mercury has a sudden situation, or recurrent intervals of disruption in his domestic or periodic contact with the Moon. This breaks up the present conditions of adaptation. With consideration and concern, Mercury can develop a deeper understanding; if he finds the agitation too disturbing, he can drop the matter.

The break up of present conditions often paves the way for the two people to begin a relationship. In the long term examples, it is more indicative of disruptive periods that became too uncomfortable to continue. Mercury withdraws from the contact; he may or may not renew a response at a later date.

A mother with Mercury sesqui-square the Moon of her daughter had an instinctive love and understanding, but the daughter's moods and feelings would periodically build up a period of disruption. Their lifetime contact vacillated both in harmony and in timing. Several times they broke up and separated in distrubance, but always renewed a responsive exchange of adaptation.

Several cases were similar, in which the woman had Mercury sesqui-square the Moon of the man. Mercury was attracted to respond to the mood and feeling of the man, but when they formed a domestic or periodic contact, it would not remain consistent.

The intervals of disruption came from their clash in instinctive processes as well as from circumstantial situations. Their interests varied, their moods varied, their home tastes varied.

Quite often, Mercury simply gives up, as the disruption hardly seems worth the effort.

The progressions show either the time when the two people meet, during a fluctuating domestic period, or a time when their situation changes to new lines or breaks up.

Mercury Inconjunct Moon

Mercury has an instinctive response to the moods and feelings of the Moon. He takes an initiative of concern and nourishment, protection and care of the Moon in a domestic or periodic contact. In time, Mercury's response disperses into other directions that need his attention. His contact with the Moon can remain firm, though it may have minor fluctuation in mood.

By progression, the inconjunct indicates a time of domestic climax in Mercury's response to the Moon, after which it disperses.

These examples are of people who had an immediate exchange of adaptation and sympatico. In most of the examples there was a long term relationship that contained a basic understanding throughout the vicissitudes of other aspects and events.

In one example, where a woman's Mercury was inconjunct a man's Moon, they met with an immediate response of mental attitude. They married and had two children. Though Mercury's attention and care was directed to the children, whenever her husband had a change of mood or feeling, she would instinctively respond to him. Due to other complexities of aspect, they divorced, but throughout the children's growing years, they maintained an understanding relationship.

In other long term examples there were periodic times when Mercury was stimulated to a penetrating response of care and solicitude for the Moon that would gradually settle into routine.

The progressions show a time when Mercury instinctively assumes the protection and care of the Moon, followed by dispersement.

Mercury North Parallel Moon

Mercury has a different life pattern than the Moon in his domestic life and day to day experience. His instinctive mental process that determines mood and feeling has been diverse.

The two come together with spontaneous circumstantial ease in a domestic environment or close periodic contact. They have a sympatico of mutual adaptation. They can blend for a period of time with an exchange of concern, protection and care.

However, the circumstances of their diversities are such that eventually Mercury is impelled to express his day to day experiences in his own familiar environment. Their contact can continue on a periodic basis at intervals.

The examples indicate that Mercury will have a close and frequent contact with the Moon over a period of time. Two women lived as neighbors with a daily contact for two years, until the Moon moved away. After that, they saw each other once or twice a year.

Another example was of a theatrical agent who had Mercury N parallel the Moon of an actor. During the two years that he worked in TV, they had contact several times a week. For many years, he was an occasional guest in Mercury's home.

The Mercury N parallel Moon progressions indicate the period of time when there is the bond of sympatico that impels Mercury into a detailed contact with the Moon. In each example, the two people had a periodic contact that was daily or weekly, over most of the time of the progression.

Though some of the examples were in a work environment, they had greater closeness when they overlapped into the home.

In one example of two women with Mercury N parallel Moon, Mercury went to work for her sister-in-law, the Moon, who was a fashion designer. Their association overlapped into both homes, where they exchanged an adaptation of sympatico in both business and personal detail of daily experience. Their home environments were highly diverse. Their mental processes were highly diverse. After the aspect moved out, they continued to see each other on occasion, but without the intimate bond of mutual protection and concern that they had during the year when they had the Mercury N parallel Moon.

As the Moon rules the general public, and the parallel includes the environment, the Mercury parallel Moon aspects have a larger number of examples where the two are in public situations. None of the Moon aspects imply business contact, but the consistency of periodic contact in a public situation can be attributed to the Moon.

Mercury South Parallel Moon

Mercury has a different life pattern than the Moon in his day to day experiences and in the instinctive mental processes that determine mood and feeling. However, they have large areas of similarity that overlap in their domestic situations and in certain mental attitudes.

They come together naturally as the result of circumstances in a common environment, usually the home, where there is a daily or periodic contact, with a certain tension.

Mercury has a diversity to the Moon in his daily pattern and attitudes, but their bond is such that he is impelled to give solicitude or protection.

Where the S parallel examples show a strong overlap in life patterns, it must be remembered that this is not a total blend; Mercury does have a life pattern to fulfill apart from the Moon.

A father had Mercury S parallel his daughter's Moon. During her childhood, Mercury had a traveling job, so his home contact with his daughter was intermittent. When the girl was 16, father's work stabilized to where he was home daily. The adolescent tempo of music, mood and feeling caused a tension at home for Mercury. His daily patterns and attitudes differed from those of the Moon, but their bond was such that he had a great tenderness for the girl.

In many cases, there is a frequent contact in the home of one or both, where they do not live together.

The progressions are more indicative of a period of time when Mercury is impelled by circumstance to care for the Moon. The Moon may have a problem of health, welfare or attitude to which Mercury responds with solicitude.

In one example, an adult son with Mercury S parallel his mother's Moon, left his home to move in with her during the winter she was ill.

In another example, a man had Mercury S parallel his wife's Moon at the time when she had an extra-marital affair. As this was not her usual pattern, she became moody, inconsistent and over-emotional. Mercury related to her with protection and understanding, to stabilize their home.

There is a fluctuation of contact and mood during the Mercury S parallel Moon progressions.

In all the S parallel aspects, Mercury can fulfill a relationship of tenderness and solicitude for the Moon, while maintaining his own domestic, or day to day patterns.

Mercury Split Parallel Moon

At some time, or over a fluctuating period of time, Mercury has a maximum contact with the Moon in an exchange of adaptation and penetrating understanding. Each has certain day to day experiences apart from each other, that are uninteresting or incomprehensible to the other.

Where they are impelled together in an exchange of care and solicitude, they can maintain a cohesive unity only in a candor of mood and feeling, and an honest exchange of their mental processes. They tend to be forced apart in direct ratio to their accumulation of small secrets, gossip and fluctuating mood.

The examples include marriage and family relationships.

There is a rare example of a friendship that is not set in the home, but where the two people met in the office periodically over many years. There was a commercial agent with Mercury split parallel the Moon of an actor whom he sustained during the lean years, and publically acclaimed during the prestigious years. They had an intense rapport of mutual solicitude.

In another example, an orthodontist had Mercury split parallel the Moon of a young girl on whose teeth he put braces. They had an intermittent contact over a three year period.

In every case, the two had some level of daily experience apart from each other.

In a husband and wife example, they both were quite indifferent to the other's daily details, as they had an exchange of mood response when they were together.

The progressions show the period of time when Mercury makes the maximum contact with the Moon in the home or in a period of detailed exchange. During this time, Mercury responds intensely to the Moon with care and solicitude, while still maintaining certain day to day experiences in a separate environment.

We have the Mercury split parallel Moon of a man to his son at the time of the child's birth.

A woman had Mercury split parallel the Moon of a man who was her husband's friend and co-worker. During the four years of the progression, they were together several times a week in the home, as both couples were socially compatible.

Another couple married at the time of this progression.

If there are secrets in their condition of health, mental attitudes, mood or feelings, the progression shows the time when they lose trust and diminish their periods of contact.

In a business exchange of a public matter, two men with Mercury split parallel the Moon signed a contract agreement. The Moon concealed a serious illness that kept him from fulfilling the terms of the contract. There was a loss of both trust and association.

The Chart Comparison on page 60 illustrates a daily exchange in domestic and work environment. It is of two men who were not related, but who accepted each other as friend and kin.

This chart comparison is not meant to illustrate a specific event, but a quality of relationship. The chart will not be progressed.

Roger and Jack met in 1958. With natal Mercury inconjunct Mercury, they had an immediate acknowledgement of each other, and communication on similar interests.

Roger was a man of 42, a Virgo, well educated, modest, and discriminating. He had been married and divorced. He owned his own home and a small local shop that sold and repaired radio and television sets.

Jack's age was 24; he was a Capricorn. He was serious, studious, and socially shy due to a hearing problem. Jack was a good worker and with Moon in Virgo, had a skilled astute mind. His main interest and passion was in electronics.

(Mercury sesqui-square Sun.) Shortly after the two men met, Roger lost his assistant and co-worker in the shop. He offered the job to Jack.

Chart Comparison in Domestic and Work Environment

	Roger to Jack		Jack to Roger
☿ Natal 17♍44 9 N.35	⚻ ☿ 1°	☿ Natal 18♒15 16 S.36	⚻ ☿ 1°
	⚼ ☉ 2°		⚻ ☉ 2°
	"☌ ☽ 5°		⚻ ☽ 3°
	△ ♄ 6°		⚹ ♄ 4°
	△ ♂ 2°		SPLIT ♇ P
	□ ♃ 4°		S. ♇ Asc
	⚻ ♅ 1°		

(Mercury inconjunct Uranus.) Roger was impressed by Jack's inventive skill. He encouraged Jack's experiments that led to improvements in the electronic equipment.

(Mercury trine Saturn.) Their ability to work together was spontaneous. Because of Jack's hearing difficulty, Roger took the phones and customer contact, but he gradually allocated the full responsibility of shop work to Jack.

(Mercury trine Mars.) There often was pressure of time commitments on the work. Both Roger and Jack found creative satisfaction in the shop accomplishments. Roger easily stimulated Jack's energy to creative goals and their business began to build up.

(Mercury square Jupiter.) Roger began to put more and more trust into Jack. He depended on him for business capacity and for understanding. At times, his demand was for more than Jack could supply.

One summer, Roger took a long trip, leaving Jack in charge. Jack was not able to handle the desk contact well, and he was untrained in bookkeeping. The summer was a financial loss.

Roger was a big enough man to understand that it was a lack of good judgement of his own in placing Jack in a difficult position. With a continuation of loyalty, he re-established both the business stability and their good will.

(Mercury conjunct Moon.) Roger became very familiar with Jack's moods and feelings. In their long period of close daily association, he could identify many of his own deep character traits in Jack's Virgo Moon mental processes.

Jack moved into Roger's house. There was no area of basic clash between them, with this strong harmonious chart comparison. They worked well together at the shop and at home. There was an intellectual and creative nourishment of each other and an exchange of protection and trust.

(Mercury inconjunct Sun.) From their first meeting, Jack related to Roger's authority and goals without reservation.

(Mercury inconjunct Moon.) Jack was responsive to Roger's mood and to his mental processes. There was no need to question the affinity and understanding they had, both in attitude and intellectual interests.

(Mercury sextile Saturn.) Jack liked his work. He accepted the opportunity eagerly, even though the initial wage was not outstanding. He and Roger built an exchange of security, both emotionally and mundanely.

(Mercury split parallel Pluto.) Jack and Roger had a maximum relationship in their cooperation for a mutual benefit and for the good of the business and customer.

It was a depth of contact not only in their hours of work together, but their long talks of their ethics, spiritual goals, deep meanings and stresses.

Jack continued schooling on electronics. His absorption in this depth of study excluded Roger.

Roger had certain community interests that were not shared by Jack. He belonged to a men's club, and to a political organization.

In their levels of cooperative experience and aspiration apart from each other, they did not build up secret or covert areas. When one wished to share a point of discussion, he did so. Generally, these certain areas were not shared due to simple lack of interest.

(Mercury S parallel Asc.) Jack and Roger neither looked alike nor had personalities that were similar. Many of their attitudes and certain personable traits were alike. Their tastes were quite similar, so they appreciated their home, their food, and each other's appearance in the same way. They borrowed each other's shirts and ties with small jokes about the other one's bad taste.

The impelling factor of their circumstances and affinity drew and held Jack to Roger.

Jack did have a need to express his own personality, physical and emotional drives apart from Roger. This fulfilled a normal course when he met a girl that he loved.

In 1962, Jack married, and moved into his own home. He went into full partnership with Roger. Their relationship deepened and endured through the years.

CHAPTER VII

Mercury in Aspect to Venus

VENUS RULES the social and emotional impulses. It is the cohabitative urge of affection and mating, or companionship, ease and pleasure.

Mercury makes contact with Venus on these matters. He has an emotional exchange with Venus in affairs of beauty, mirth, pleasure, music, parties, games, satisfaction, harmony, ease, and graciousness.

The Mercury-Venus aspects are most pronounced in friendship, appreciation and love.

The discordant effects are in unrequited love, hurt feelings, heartbreak, distaste and in disturbing emotional attachments.

The Mercury-Venus parallel implies an intense exchange in which the feelings that the two people have contain a range of similarities and differences. It indicates their emotional relationship within the larger context of their individual social lives.

Mercury Conjunct Venus

Mercury has a familiarity with the social and emotional impulses of Venus. He understands or identifies from his own personal experience how Venus feels about a certain situation. They will share, or experience at the same time, a similar emotional situation, or social affairs in common. They have a love or friendship with each other or for the same person.

As the conjunction is neutral, Mercury may acknowledge the beauty, graciousness or harmony of Venus without emotional involvement.

In all the examples, Mercury has an appreciation or admiration of Venus. There are a few marriages, friendships, and family ties. Approximately one out of three examples are of people who live together for a period of time, but the aspect does not in itself imply intensity or depth.

In a number of examples, Mercury has an intermittent social contact with Venus, in which he warmly regards the attractiveness, the artistry or musical ability, or striking charm of Venus.

Most of the couples have friends in common.

In one example, the Mercury of an orthodontist was conjunct the Venus of a young girl. He understood well her feelings about the treatment. They shared the experience of increasing her beauty.

In another case, two women neighbors had Mercury conjunct Venus. Mercury admired Venus as a strikingly handsome woman. They shared a love for their children who played together. Mercury understood and identified with the emotional turmoil of Venus at the loss of a baby, as Mercury had experienced a similar loss. Outside of attending a few concerts and neighborhood functions together, the two women had no particular depth of friendship for each other.

The progressions show more vividly a time when the two have a common emotional experience or situation. The man who raped the little girl (refer to Mercury split parallel Mercury) had Mercury conjunct her Venus at the time.

In other examples, there is an increased social life for the two people; Mercury has a love affair similar to the one that Venus had; or their contact changes from business to a social exchange that is pleasing to both.

Mercury Opposition Venus

Mercury has in common with Venus a love of beauty, music, art, pleasure or ease, affection or companionship. Where they reach an understanding or identification on this matter in common, their emotional natures, and manner of expressing love does not fully blend.

The relationship can remain constant when held on the level of their inter-capacity. When the need of one or the other for emotional expression develops to the point of impasse, they withdraw from each other. As Venus is benefic, it tends to keep events serene, and often returns the two to the same degree of esteem as before.

In some of these examples, there is an intermittent contact on social and affectional levels where Mercury is quite appreciative of Venus. In other examples, there is long term marriage and family co-habitation.

A man and woman with Mercury opposition Venus, (who also had Mercury split parallel Venus) had a life time marriage that was rich in warmth and companionship. There were periods when one or the other would withdraw with hurt feelings or misunderstandings. There were several times when Mercury, or Venus, was attracted to another, but they always returned to a mutual appreciation of each other.

Several other marriages slowly built an inflexible emotional struggle of incompatibility over periods ranging from 6 months to 15 years. There was a breakdown in understanding, hurt feelings or heartbreak and a period of withdrawal.

Circumstances parted the two people in many of the examples. They were separated by distance or one of the two married another person. In four examples, Mercury was heartbroken from the death of his beloved Venus, who was father, brother and friend.

The progressions show the period of time when the two strike a certain level of companionship and affection that tends to remain consistent; they seem unable to either consummate or break up the level. There is some misunderstanding and inability to adapt between them. The circumstances of their lives holds them apart, while their emotional regard for each other holds them together.

Several friendship-romances had a companionable warmth, but when Venus was ready to love, Mercury was busy; when Mercury was amiable and warm, Venus was involved elsewhere.

In all of the Mercury opposition Venus examples, the two people have some social life together, family or friends in common. There are two exceptions, in doctor-patient relationships.

Mercury Square Venus

Mercury is strongly attracted to an appealing quality of Venus. He admires the beauty, charm, artistry, poise or sweetness of Venus.

However, there is an emotional incompatibility between the two, or circumstantial obstacles to their expression of love, in mating or in friendship.

A cheerful, gentle and patient Mercury can draw out a depth of companionship and harmony with Venus, but an over emotional and touchy Mercury builds distaste and misunderstanding.

There are few family relationships here. The Mercury square Venus is found mainly in friendships, marriage and romance, and in people who meet with some emotional ambivalence.

In many examples, Mercury seems to feel that he could love Venus if Venus were different in some way, or if circumstances were different. A woman with Mercury square Venus to a man, found him charming and amiable, but untrustworthy and amoral.

Another woman with Mercury square Venus to a man began a romance, but found in him a weakness that she thought unmanly.

A man with Mercury square Venus to a woman was charmed by her appealing femininity but soon found her sickeningly sweet.

In other cases, the love, or the manner of expressing love, is incompatible between the two.

In over a third of the examples, an emotionally mature Mercury overcame obstables of misunderstanding with patient trust and appreciation of Venus. These examples are outstanding for the exchange of warmth, tenderness and love between the two, in cases of friendship, marriage and kinship.

The Mercury square Venus progressions are more indicative of circumstantial obstacles. Mercury may be attracted to Venus at a time when one is married to another. Mercury may be involved in some emotional turmoil that causes incompatibility.

In some cases, there is a crisis situation that causes emotional impact. Several examples are of men and women who had a secret love affair with each other, though married to someone else.

In one case of Mercury square Venus, the two women were close friends. Mercury's child was molested by Venus' husband at the time of the progression, causing emotional crisis to both.

In another example, a young child had Mercury square his mother's Venus. He was having problems in social adjustment in school that caused her disturbance.

In all cases of Mercury square Venus there is some social exchange, or social life in common.

Mercury Trine Venus

Mercury has a spontaneous appreciation or admiration of Venus for a quality of character or personality. Mercury loves Venus, as a friend, companion or mate. As the Mercury trine Venus aspect is placid in nature, that love may be no greater than a casual pleasure in the company of Venus. When other aspects intensify the contact between the two, Mercury's love can have endurance and depth.

Where the examples do have long term friendships and family relationships, there are surprisingly few enduring romances and marriages. When events do not go easily and harmoniously between the two, Mercury withdraws from Venus, as his passions and efforts go elsewhere.

A woman with Mercury trine a man's Venus fell deeply in love with him. They had an affair for a year that became increasingly tempestuous the second year. Her moody demands became distasteful to the man; her urge that they marry was disagreeable to him and they separated.

In another example quite similar, the two did marry. The Mercury woman made only as much effort to please Venus as happened to suit her pleasure, and they eventually drifted apart.

In the close friendships, Mercury gives a measure of joy and appreciation to Venus during an intermittent contact that is both sociable and affectionate. There are few examples of long endurance in cohabitation outside of the family relationships.

The progressions are more encouraging, as they show a time when Mercury is spontaneously drawn to Venus in love and admiration. These people often began love affairs, marriage or cohabitation.

Two friends who had for several years been separated in enmity, were reunited with the Mercury trine Venus aspect.

In the majority of Mercury trine Venus aspects, there is some social life in common.

Mercury Sextile Venus

Mercury at some time has the opportunity to demonstrate his love, appreciation and admiration of Venus. At that time the two are drawn together easily and the way is made clear for Mercury to express himself. If Mercury does not make that effort, the time passes and the warm exchange of harmony is lost.

The examples include all types of contact, in business, school, family and friends and love affairs. The chance seems obvious enough, for there is a time when the two live, work, or play together. As long as Mercury gives his approval freely, Venus is amiable, pliable, helpful or in some way gives mercury ease or pleasure.

If or when Mercury chooses not to extend his yielding companionship, the two drift apart, often with hurt feelings and misunderstanding.

A mother with Mercury sextile her daughter's Venus, gave her approval, appreciation and gratitude during her adolescent years. The daughter was a helpful and considerate Pisces who found innumerable ways to comfort and please her mother. They built a lifetime bond of love during these years.

Two men with Mercury sextile Venus were friends and business co-workers for years. Mercury gave Venus approval and camaraderie and Venus often took the heavy load from Mercury's shoulders, with financial assistance, or helpfulness in contact and business arrangements. When Mercury's business expansion took him into changing goals, he had less time or interest in Venus. It was hard for Venus to understand this, and he felt hurt at the diminishing friendship.

A man with Mercury sextile the Venus of a beautiful woman was highly attracted to her. Though she was receptive to him, his business was at a pressure peak of output and he did not make the effort of pursuit. Later, his interest renewed, but she was engaged to another.

The progressions show a time when the two come together on an emotional or social matter that can be beneficial to Mercury if he makes the effort.

Several marriages began; several long term marriages and friendships were reaffirmed at this time with greater appreciation. In other cases, Mercury did not extend himself to give any approval, and the chance was lost in time.

Where the Mercury sextile Venus is a pleasant exchange, it does not withstand stress or changes. It will not insure a relationship unless backed up by stronger aspects.

Mercury Semi-sextile Venus

Mercury has a vivid spark of interest in the attractive appeal of Venus. With an emotional response from Venus, Mercury is quite willing to put the effort into developing a warm relationship.

In one example of Mercury semi-sextile Venus, the man said, "She's a beautiful woman with an exceptional quality, but there's something lacking in her feelings."

All the cases seem to show some spark of misunderstanding or hurt feelings along with the interest, until both persons make the effort to attain harmony.

There is cohabitation and friendships among the examples, where other aspects must be considered for substance in contact.

The progressions show the time when Mercury's spark of interest can be fanned into love or appreciation, but with a great deal of effort.

In several examples, Mercury made a luke-warm attempt to begin a romance. A Mercury mother had a period of concern about her son's emotional stability. In a case of two neighbor women, Mercury had a flare up of touchy feelings and stopped speaking to Venus.

The friendships flourished nicely with the Mercury semi-sextile Venus.

Mercury Semi-square Venus

Mercury's contact with Venus is intermittently pleasing and irritating. Where Mercury does love or appreciate Venus, he finds it hard to understand the emotional attitude that Venus has. Mercury is occasionally critical, as he has some worry about Venus' feelings, and the harmony of their relationship vacillates. Mercury can handle these slight changes with patience and firmness as the emotional exchange either stabilizes or loses importance to him.

A relationship can continue in this tone for a lifetime. A practical, ambitious brother with Mercury semi-square his sister's Venus, admired her charming incompetence, but found it irritating when she was late or forgot a commitment to him. When she had a series of erratic romances, Mercury was critical of both her choice and her feelings. Later, Venus married a man with whom Mercury got along very well and the brother-sister love stabilized nicely. In maturity, Mercury's own family took his greater interest. His occasional contact with his sister often sparked a flare of impatience, but their love continued through their lifetime.

The progressions show the time when Mercury's attraction or concern for Venus is mixed with impatience. A woman had Mercury semi-square her daughter-in-law's Venus at a time when the young people were having marital problems. Though the mother had anxiety, in her love for them both, she tried to avoid involvement. She privately felt that the daughter-in-law was being overly dramatic in her emotions.

The young people did smooth out their problems, and the love of the mother and daughter-in-law remained firm.

The majority of the Mercury semi-square Venus aspects are between people who had close and long term relationships.

Mercury Sesqui-square Venus

Mercury has a sudden situation, or recurrent intervals of emotional disruption with Venus that tend to break up the existing relationship. Mercury must make the decision of drawing on his own serenity to overcome his agitation, if he is to appreciate Venus, and continue their relationship.

It may be Mercury's own emotional agitation that is disruptive, though in most cases, it is the situation in which both are involved.

An actor's Mercury in sesqui-square to his agent's Venus always had contact with him in the sudden disruption of casting calls. Mercury had to maintain his serenity to answer the interviews and work calls with poise.

A woman with Mercury sesqui-square her employer's Venus admired both his business capacity and personal charm as their work relationship overlapped into social affairs together. When he fell in love with her, Mercury was not emotionally prepared to carry through a steady romance and the disruption broke up their contact into agitating intervals until it stopped entirely.

In another example, a woman with Mercury sesqui-square her girl friend's Venus was hurt, rejected and touchy when Venus fell in love with a man who had dated them both. Mercury made every effort to be understanding and their contact continued, but with decreasing intervals of harmony.

In the progressions, the two people meet with a magnetic impact where Mercury finds Venus quite upsetting to his way of life. In one case, the two met, had a brief and dynamic romance and separated. Mercury felt that this woman was entirely too unconventional for his ambitions of social position.

In a similar case, the two married but at the cost of both divorcing their present mates during a period of highly agitating change.

A long term marriage went through a rocky period during the Mercury sesqui-square Venus aspect, as the wife Venus was spending her Mercury husband's money at a rate that was quite disruptive.

The aspect can be found in an enduring relationship but not without some periods of change and agitation.

Mercury Inconjunct Venus

Mercury is strongly attracted to Venus for a quality or appeal that Venus has. He takes an initial interest in developing a social or emotional relationship. In time, his expression of love is directed elsewhere but he can retain a firm esteem or camaraderie for Venus.

In a marriage based on Mercury inconjunct Venus, Mercury's greater expression of love may be directed to the children, or in the case of a male Mercury, to a friend, while maintaining the compatibility of the marriage. When the love does go to another person, there is in some cases, a period of infidelity through which the marriage survives.

In a case of a parent with Mercury inconjunct his child's Venus, the parent may express his adult love needs with an adult, while continuing a lifetime adoration for the child.

There are four deaths among the aspects, twice of Mercury in the natal examples, twice of Venus in the progressions. One of these was of Mercury inconjunct the Venus of his murderer. These examples are illustrative of the emotional impact that sometimes accompanies the aspect. Of course, the much heavier aspects are more indicative of the seriousness of death.

By progression, the Mercury inconjunct Venus indicates the time when Mercury is emotionally attracted to Venus, but the contact is seldom consummated. In several cases, a woman met her future husband when her Mercury was inconjunct his Venus. Though she was interested immediately, it was not until the aspect moved out of orb that they began their romance.

In other examples, the two maintained a placid and intermittent friendship.

Occasionally, there is the impact of Mercury or Venus having some emotional upset. In several examples, Mercury fell in love with another, while maintaining his bonds of warmth and esteem for Venus.

Though there are examples of separation, usually in the sudden romantic affairs, the inconjunct is not always separative by progression. Other aspects must be considered.

Mercury North Parallel Venus

Mercury has different emotional patterns than Venus. Their social environment and experiences have been diverse, with some few similarities.

They come together spontaneously, in a social exchange. Mercury is attracted to some quality or appeal of Venus. This hits a responsive cord in both persons. They can blend for a period of time in affairs of beauty, harmony, pleasure or love. Mercury will develop this as long as it fits in with his primary life pattern.

However, the circumstances of their diversities are such that Mercury is impelled to find his love, or social or emotional expression on his own customary level.

There is seldom a total blend of the expression of love between these two people. They feel that they are unable to change their emotional or cohabitative situation. Whether Mercury regards Venus with feelings of respect, admiration, affection or longing, his emotional needs and conditioning have a difficult diversity to those of Venus.

In several similar examples, the two met at a party. In the best romantic tradition, their gaze met, they left the side of the person they were with as they were drawn together. There was a helpless, wordless yielding of falling in love.

In one of these cases, Mercury was securely married to another. In another, Mercury's social world was such that Venus would be totally out of place. In a third, Mercury had a period of turmoil before deciding he was in love with another girl.

One marriage lasted eight months. In that time, Mercury had multiple affairs with others.

There are few examples where the two people did not meet in a social situation. In one, a woman had Mercury N parallel

Venus to her beautician, whom she saw periodically with appreciative harmony.

The progressions show the period of time in which Mercury has varying degrees of exchange with Venus, from mild admiration, to warm camaraderie, to intense attraction. In the majority of examples, Mercury is emotionally involved with someone else. Usually, Mercury is in love, or having a love affair. In other examples, Mercury is going through an emotional crisis with their beloved friend or mate, parent or child, apart from Venus.

The relationship between Mercury and Venus can be serene except that in some way Mercury's feelings about the third person reflects on his contact with Venus.

This can be as simple as the three year old boy with Mercury N parallel his mother's Venus. He found his first friend, an inseparable companion, whom his mother thought of as a bratty little monster.

In another example, two women who had been friends since childhood had the Mercury N parallel Venus progression at a time when their contact was limited to an infrequent social visit. After a disastrous love-affair, Mercury had an emotional breakdown and was admitted to a mental hospital. Venus was shocked and heart broken at what she felt was the loss of her friend.

Whether this is a time of harmony and satisfaction for Mercury, or a time of emotional crisis, the progressed aspect is not totally separative to Mercury and Venus except to the extent that Mercury must express a certain emotional impact apart from Venus.

In the natal and progressed examples, there are a few rare cases where the two people changed their patterns in order to fully interrelate or cohabitate. These people had an exceptional depth of love and understanding.

Mercury South Parallel Venus

Mercury's basic pattern of social and emotional expression differs from that of Venus, however, there are large areas of similarity that overlap.

They come together as a natural result of circumstances into a common social environment. Mercury is impelled to respond emotionally to Venus. The situation or condition of their contact has a tension that affects them in diverse ways.

These people often live together, or associate in friendship for many years. Though Mercury's feelings are directed toward Venus, the contact is not always smooth. It indicates usually an emotional tension.

In one example, a couple lived together for 20 years. It was to all appearances a serene marriage. Mercury was a dignified and taciturn Scorpio man, who concealed his depths of feeling. After his death, Venus confessed privately that it had been a cold relationship, in which she had become increasingly discontent.

In a business relationship, two men with Mercury S parallel Venus had been friends since their youth. They were both musicians who played in a combo, and had the experience together of night club work. Mercury's admiration for Venus was ambivalent, as he also had a prejudicial distaste for their racial difference. They had sharp disagreements on certain moral issues, and their private social life seldom overlapped.

The progressions indicate a time when Mercury is impelled to respond to Venus in emotional tension.

The examples include two marriages under duress because Venus was pregnant.

A woman with Mercury S parallel the Venus of a man, had a business relationship with him that she made every effort to change into a personal contact, but was rejected.

A man with Mercury S parallel the Venus of his sister took her into his family home with sympathy after she'd been sexually attacked.

In one example, a woman had Mercury S parallel the Venus of a man for three years. At the beginning of that time, he seduced her; within two months she went to live with him. They undoubtedly loved each other, and it was on the whole a happy three years. However, his temperament was not suited to monogamy and he had many other affairs. There was some tension in that Mercury wanted to get married. When the aspect moved out, they did marry.

In all the Mercury S parallel aspects, Mercury responds to Venus, but on a level of ambivalence.

The aspect is unifying to the extent that Mercury must fulfill a meaningful relationship of love or social unity with Venus while maintaining his own primary emotional pattern.

Mercury Split Parallel Venus

Mercury's maximum contact with Venus is on a matter of beauty, music, art, social exchange, or love.

Both have a social, emotional or aesthetic level apart from each other, that is uninteresting or incomprehensible.

They can blend in a unity of love that is familial, fraternal or erotic to the extent of their equality in emotional maturity. Where there is honesty and appreciation, the aspect impels the two together. Where there are hidden feelings, secret misunderstandings or amorality, the two are forced apart.

The examples are evenly divided in the couples that blend together or separate. In all cases, the two have family or friends in common. In many cases, they both love the same person, who may be a child, or a parent, or kin to one and friend or lover to the other.

There are more marriages in the split parallel than in any other Mercury-Venus aspect. Usually these are good, cohesive marriages, when based on emotional candor and friendship. When the marriage, friendship or kinship is based on pleasure seeking, personal gratification, or moral intemperance, the two part with distaste, hurt feelings and heartbreak.

As an example of this, one woman with Mercury split parallel the Venus of her brother had for a lifetime a familial love and loyalty for him but a distaste for his alcoholism and his amoral misadventures.

The progressions imply the time of maximum emotional exchange. As the parallel brings in the whole structure of environment, the relationship includes those portions of the life pattern that each has apart from the other.

There was a time when a mother and daughter with Mercury split parallel Venus clung to each other at the heartbreak of the death of their husband-father.

A woman with Mercury split parallel her brother-in-law's Venus gave him admiration and liking, but when she found that he was secretly unfaithful to her sister, she withdrew approval. They concealed their growing distaste for each other in a diminishing contact.

In another case, a man had Mercury split parallel the Venus of an older woman, whom he adored. Though he was about to marry at the time of this progression, he and the old lady were closer than at any time in their lives. Perhaps the old lady knew that she was soon to die, for the honesty of her devotion was exchanged in an open heart-to-heart contact.

A Mercury woman had split parallel Venus to her son's girl friend. The two women blended nicely in similar social and artistic tastes and the mother was delighted when her son married the girl.

There are examples of intense romances, marriage, and friendships between Mercury and Venus. In some examples, there are milder expressions of harmony or camaraderie among people who never exceeded this maximum point reached at the time of the progression.

It is a complex aspect, that brings the whole of two separate life patterns together into common experience.

The Chart Comparison on page 78 illustrates a long term cohabitation in love and marriage. The couple are Dave and Margaret.

Margaret was a red haired Irish girl, with the quality of endurance and tenacity of her Cancer Sun-sign, the courage and initiative of Aries Asc, the strength to accept responsibility of Saturn in the first house.

Davie was a young Frenchman, filled with the vision and dreams of his Pisces Sun-sign, to the adventure of colonization, with Jupiter conjunct Neptune in his 7th house.

They met in 1894, when Margaret was 16 years old, Dave was 25. Dave was out of the Spanish-American war and drawing a government pension of $8.00 a month.

Chart Comparison Illustrating a Long Term Cohabitation

	Margaret to Dave		Dave to Margaret
Natal ☿ 6 ♌ 51 20 N.00	△ ☽ 1° ⚻ ♀ 2° ☌ ♂ 8° △ ♄ 8° △ ♃ 4°	Natal ☿ 29 ♒ 00 15 S.00	☌ ☽ 9° △ ♀ 9° ⚻ ☉ 1° ☍ ♆ 2°
Progressed 1895 ☿ 4 ♍ 10 8 N.00	Split P̲ ☿ Split P̲ ☉ △ ♀	Progressed 1895 ☿ 18 ♓ 00 8 S.00	Split P̲ ☿ ⚺ ☽ □ ♀

(Mercury square Venus, progressed.) Dave fell in love with the tall, full-bodied girl and pressed his courtship against the obstacle of her initial reserve and the family disapproval.

(Mercury trine Venus, progressed.) Margaret responded with emotion that spontaneously increased.

(Mercury semi-sextile Moon, progressed.) Dave's spark of interest in building a home and marriage with Margaret grew into reality.

(Mercury split parallel Mercury, progressed.) Their closeness increased to a maximum of understanding and identification of each other. They shared their thoughts, and planned their future together.

(Mercury split parallel Sun, progressed.) Dave and Margaret found a unity in their integral character of goals and meanings and reached a maximum of vital relationship in their marriage.

In 1912, Dave and Margaret homesteaded 320 acres of land on the Canadian prairie. They had $120.00 in cash money, ten head of good stock, a buggy and used lumber to build their home and barns themselves. They also had a dream, their courage and faith in their future together.

The years ahead were hard years. There was illness, poverty and desperately hard work. Thirteen children were born.

Dave worked as a laborer to get the money for grain to plant. Margaret sold butter and eggs and kept a garden. They fed their family from the products of the earth, butchered and hung their own meat in a smoke cellar. Margaret hand carded and made eight mattresses, quilts, bedding and the children's clothes.

At night, they would read by oil lamp out of the Bible and mail order catalogues, their only books.

They took more land and made it grow. The children were educated in a country school house where Dave worked as janitor, while Margaret took the school teachers to board.

(Mercury trine Moon, conjunct Moon, natal.) They were married for 70 years, with a full blend of response to mood and feeling, in the care and solicitude, protection and trust of each other.

(Mercury inconjunct Venus, trine Venus, natal.) The full range of their social and emotional exchange was experienced in those years together. With the two natal Venus aspects, and the two progressed Venus aspects, their love was radiant and enduring.

(Mercury conjunct Mars, trine Saturn, trine Jupiter, natal.) The accomplishment of their labors and responsibility slowly built the security of a homestead without debt. When all the children were still home, they had 1500 acres of rich farmland, 70 head of cattle, 75 head of horses, an Angus bull, a Percheron stud, 60 pigs, chickens and geese. They worked from 4 a.m. to 10 p.m.

(Mercury opposition Neptune, natal.) Dave and Margaret had together a Utopian dream. They achieved that dream, but did it through a reality that often was desperately harsh and bitter. There were disappointments and bad debts. One summer, the locusts came in a cloud that blackened the daylight and covered the land with devastation. Margaret and Dave ran through the fields setting them on fire, sobbing and cursing.

Other years, there was hail, or rust on the grain.

There were the tragic losses of the deaths of two of their young children from disease, and two of their adult children of cancer and of arthritis.

In 1964, Dave died at the age of 95. Margaret followed him the next year. They left a dynastic heritage of pride, of the dignity of labor and the endurance of courage.

CHAPTER VIII

Mercury in Aspect to Mars

MARS RULES THE ENERGY. It expresses through construction or destruction, aggression, combat and sex. The trauma of Mars is in force, hostility or injury.

Mercury makes contact with Mars on these matters. The two people engage in a common or conflicting increased expenditure of energy in action. The passions may be directed to the physical satisfactions of sex, eating and drinking, athletic or social competition. The relationship can be highly constructive.

The negative use of Mars energy implies inordinate or covert lust, hate, struggle and all manner of injury and violence.

The Mercury-Mars parallels show the intense similarities and differences in desire, trauma and the use of energy between the two people, within the larger context of their individual aggression patterns.

Mercury Conjunct Mars

Mercury has a familiarity with the manner in which Mars expresses his desire patterns. He understands or identifies from his own experience with the wounds, the trauma or the aggressions of Mars. They will share association at a time when Mars is expending a great deal of energy, or involved in some Martian pursuit such as liquor, lust, anger or injury.

As the conjunction is neutral, Mercury can acknowledge Mars impartially.

In an example typical of the impartial identification of Mercury toward Mars, two girls worked together in a busy office. Mars had an attack of peritonitis that called for emergency surgery. Mercury was no stranger to hospitals and she visited Mars with an understanding attempt at helpfulness. She said, "Gee my heart just bleeds for Kathy, with her stuck there for a month, but there's just not much I can do."

In the various cases, Mars had surgery or a bloody accident, was an alcoholic, was uncomfortably lusty or at the least, was restless and desirous of change. Many of the examples show Mars in an aggressive position toward their work output, or in a Martian occupation, such as doctor, producer, sculptor, mechanic or soldier.

A very few of the examples are of people who had a sexual exchange that was excessively lusty for both. In other male-female examples, there was no sexual excitation.

There is only one example of injury between the two, when a Mars man tried to kill his Mercury wife. In this case, she also had Mercury N parallel his Mars, and he had Mercury square her Mars.

The progressions held less trauma and more construction. Mars in these examples, was very busy in creative work, healing, sculpturing, production and social restlessness. Mercury took a more or less sympathetic role as friend or confidant but without direct involvement.

Mercury Opposition Mars

Mercury has in common with Mars a similar construction, or a desire for the same goal or a similar way of expending his energy and aggression.

Though they reach an understanding and identification on an action, passion or trauma, their energy and desire patterns do not blend. They may reach a harsh exchange of pique, criticism, resentment, provocation or anger, and they both withdraw until they can make contact again on their level of agreement.

Where the two do not stimulate each other to conflict, they are nonetheless separated periodically by their sex or energy patterns, or by accident, injury or rejection.

There are several long term friendships in the examples where the two shared work or social action. There was an occasional critical disagreement, or a trauma or minor injury to Mercury or Mars.

The longest example of a relationship was the most traumatic. This was a friendship of two women who had, among their aspects, a Mercury opposition Mars. The year that they met Mercury had a bloody accident in cutting her hand, and a car accident. At this time Mercury was working hard. Mars was having an active sex life, being newly married.

They never lived together for longer than a few months; circumstances parted them for months to years at a time. Though there was the rare criticism, the two never argued or parted in any emotion stronger than pique.

Over a period of 22 years Mercury and Mars had various episodes of trauma separately, within a few years of each other. Mercury had a baby bleed to death at birth, Mars had a baby stillborn. Mercury fell through a glass shower door that slashed her back; Mars was in a fiery car accident in which she suffered 720″ of second and third degree burns. They were finally parted by Mars' untimely death by accident at the age of 44.

There are three violent deaths in the examples. In one, Mars died in a car crash; in one, Mars was a suicide; in one, Mercury was shot to death. In these three cases, Mercury-Mars were companions and not involved in any other Mars pursuit at the time, other than their full expenditure of energy or aggression.

Throughout the Mercury-Mars aspects there will be examples of violence. It must be emphasized here and now, that this aspect does not *cause* the violence. It maps the existence of a dynamic force of energy that is given to be used. When the individual makes good use of that Mars energy force in construction, creation, sex, or action, it cannot accumulate into trauma.

When there is a question regarding the Mercury-Mars possibility of injury, careful attention must be given to the radical and progressed charts of both persons for bad health or accident proclivities. It can be suggested that they take constructive measures in their manner of expressing energy or hostility.

No energy can be negated; it must be directed into channels that are personally and socially acceptable.

Many times the aspect is constructive. A doctor with Mercury opposition Mars to a youngster worked with him periodically for three years of physical therapy after the boy had polio.

A fashion designer had Mercury opposition the Mars of her production cutter. They often withdrew from each other in sharp disagreement, but returned to work again on a common level of construction, in which they had understanding and identification.

A woman with the aspect to her son had frequent short periods of separation from him. He was an ambitious and capable child actor from the age of four. The mother understood and identified with his constructive efforts, but had little interest in the tedious hours on the set. An employee accompanied him to work while mother pursued her own enthusiasms.

In some examples, the two people clash in a mutual identification of hostility patterns. In other cases, the understanding and identification is in their passions. The sexual exchange is highly satisfactory in itself, but we have few examples of enduring sexual relationship. The energy and desire patterns do not blend in full.

The progressions follow the same pattern but with less trauma. The two people meet in a common desire or action, clash and withdraw for a shorter or longer period of time.

Mercury Square Mars

Mercury is strongly drawn to the vitality, courage, aggression or sexuality of Mars. There is an incompatibility in the way the two express their aggressions or energy patterns that creates a period of conflict or crisis.

When Mercury is open, positive and direct, he will demand action from Mars, with a constructive criticism.

When Mercury has hostility, guilt, competitiveness, lust or power conflicts, he will act in so destructive a manner to Mars as to leave nothing of value between them.

In most of the examples between the two sexes, Mercury is sexually aroused by Mars, but there is either incompatibility in their sex patterns or there are social or circumstantial obstacles to coitus. One or the other may be married, pregnant, homosexual, virginal, impotent or promiscuous. Often there is a critical rejection of one by the other.

In some cases the couples made the effort to enjoyable adjustment. There are a few long term marriages and many short-lived affairs that were sexually satisfactory. The marriages were most enduring in the cases where the two people had the need to express varying degrees of masochism or excessive lust together. In one example, the two people outlived their passions and spent a happy old age together, wrangling and nagging each other with a companionable familiarity.

The celibate relationships probably were more harmonious. There are a number of fine friendships where the two expressed their sex and their energy drives elsewhere, but who could work and play together. In these examples Mercury had a mature, courageous, straightforward and constructive attitude. His conflicts with Mars were minor, because he met them openly.

In contrast, the greatest potential to violence is in the Mercury square Mars. Out of 57 examples, there are four cases where a third person was killed, a child, a husband, and two relatives. In all four cases Mercury and Mars had to act together for the funeral and public arrangements at a time of shock.

In two cases Mercury died violently, by suicide and by murder. In one case Mars was in a drowning accident.

There are four examples of attack. In one, Mercury tried to kill Mars. A Mercury mother protected her Mars daughter from sexual assault. The other two cases were also sexual attacks, once by Mercury, once by Mars.

In three cases Mars had an abortion of Mercury's child.

In five cases Mercury or Mars was injured in an accident. Mercury shows hostility, resentment, and rejection. He may demand action of Mars. One couple married under extreme obstacles; the girl Mercury ran away from home and the boy Mars lost his job.

The crisis can certainly be overcome by constructive effort. The Mercury square Mars is not desperately prohibitive to a harmonious relationship. It makes a decided directive toward an honest and energetic exchange of energy to attain positive value.

Mercury Trine Mars

Mercury desires some action or response from Mars. He spontaneously arouses Mars' energy for work, social games, sex, athletic or business cooperation. With little or no effort Mercury can get what he wants from Mars, but without satisfaction to the depth of his passions.

When Mercury directs his full energy of honest, courageous and direct impetus toward Mars, he derives the full benefit of a constructive relationship.

The Mercury who takes undue advantage with no commensurate effort loses the benefit.

There are a few examples of trauma that involve other aspects as well as the Mercury trine Mars, as in the case of a couple whose adult son committed suicide. In three cases Mercury was injured, causing shock and effort to Mars. In two cases Mars was killed after living for 20 and 30 years with Mercury.

The majority of examples are harmonious. In every case there is some period of tension or excitement. This can be handled constructively with a little cooperative effort that Mercury can elicit with Mars. That period of tension may be in their sexual attraction, a dispute or disagreement, an element of jealousy or competitiveness, a demand, a temper, or a trauma.

There is a vigor, a stimulation between the two people. The sexual exchange is dynamically passionate and satisfactory in both the heterosexual and homosexual examples.

If there is any hostility or disappointment it is because one or the other cannot supply the demand for depth of passion. The two can easily reconcile their differences with a little effort.

Because their circumstances are so propitious there is an occasional love affair that is gone into so easily, it has repercussions of guilt and resentment.

There are long term marriages and friendships, shorter love affairs and business contacts.

The progressions have the same tone, showing the time when the two come together spontaneously. There were several dynamic attractions, where Mercury began a love affair with Mars that lasted a shorter time than the progression.

In a long term marriage, Mars had a slight stroke at this time. Mercury put her energy and courage into taking care of him.

In an example of two friends, Mercury gave Mars a job that he handled so well it was beneficial to both.

In one case a marriage dissolved because Mercury made no effort to sexually or socially reach agreement with his Mars wife.

With the Mercury trine Mars aspect, the chance for a constructive relationship comes so easily that Mercury seldom puts any more effort into the matter than circumstances call for.

Mercury Sextile Mars

Mercury has the ability to build a constructive relationship with Mars. At a certain time, the two will be drawn together easily and they have the chance to expend their energy together, in work, sex, social or athletic games, or business cooperation.

Mercury has the choice of how much endurance the association will have. With decision and initiative he can easily arouse Mars' energy to whatever he desires.

When he takes no effort, the chance is lost in the passage of time.

The sexual contacts here were passionate with little endurance. There are few marriages in the examples. There are varying degrees of friendship, often with a measure of hostility or disagreement involved. Mercury at some time has sparks fly, or rejects Mars. The contact does not seem strong enough to withstand any trauma.

In one example, a young man had Mercury sextile the Mars of the girl he was in love with. When her roommate moved out he had the opportunity presented to move in with her. Within a few months he took her for granted. His indifferent attentions did not wear well, so another more amorous suitor took the girl away. Mercury objected to this, but his chance was gone and so was his girl.

In other examples, Mercury flatly rejected Mars attentions, sexually or socially.

In some cases there is an intermittent long term association where the two occasionally expend their energy together but do not press any advantage to deeper relationship.

There are several examples when Mercury took the opportunity, not of construction, but of sheer destruction. In one case Mercury sexually attacked Mars; in one case Mercury shot Mars.

The progressions show the time when Mercury has the chance to express his desires with Mars. There was an example of rape. There were several short affairs, in one, a woman Mercury was in love with Mars, but did not make a sufficient effort to hold his attention. In the other a man Mercury seduced Mars on one occasion, and rejected her.

Other Mercury sextile Mars combinations began cooperative relationships at this time by working together, in either a social or business capacity.

Mercury Semi-sextile Mars

Mercury has a vivid spark of interest in Mars, and with his cooperation they can develop a common desire, for each other or for an action expenditure together.

This is a stimulating aspect, without the trauma of the heavier Mercury-Mars aspects. In the examples of long term relationship the two often had a business or social exchange that was energetic and buoyant.

There were marriages and love affairs that were satisfactory. (Other aspects have to be considered for a complete picture.) In some cases the Mercury semi-sextile Mars simply indicates the attraction between two people, about which they did nothing.

The progressions are indicative of the specific time when Mercury's interest is alerted. This may be physical or mental stimulation. If there is no response aspect from Mars, the matter falls flat, but usually there is some degree of action, or energy expended.

A man with Mercury semi-sextile the Mars of his young son joined a hiking club with the boy; during that year they companionably had a number of camping trips together.

Mercury Semi-square Mars

Mercury's contact with Mars had intermittent periods of irritation and criticism. He does not fully understand the way in which Mars expresses his aggressions, or he may desire some action from Mars that is not forthcoming. There is occasional friction between the two, that causes Mercury some worry or anxiety.

Mercury at best may be patient and firm, to let the matter develop or diminish.

In many of the examples even when Mercury loves Mars, as a friend, child, lover or acquaintance, he has flashes of impatience at the way Mars does something that Mercury doesn't understand.

A Mars child had a dominant aggressive energy pattern. A Mars girl friend broke dates unexpectedly. A Mars boy friend had periods of bad temper. A Mars husband had occasional affairs away from home. When Mercury accepted these vacillations with patient firmness, the contact remained close. But when Mercury was critical or worrisome the relationship became highly strained.

In some examples Mercury dismissed Mars as an irritating bore. There was a simple clash, and rejection.

The progressed Mercury semi-square Mars puts a strain on a long term marriage that is critical. Friendships tend to become intermittent at this time, and vacillate greatly between their former closeness and the present time friction.

There were several social exchanges in which Mercury took an active interest in the activities of Mars as a friend and confidant.

Romances were begun, with the excitement of sexual attraction.

The only trauma in the examples was in a long term marriage that was going badly. The wife, with Mercury semi-sextile Mars to her husband took the initiative of telling him to get out. They were together again within two months.

Mercury Sesqui-square Mars

Mercury has a sudden trauma, or disruptive intervals of change with Mars. This breaks up the existing conditions of their relationship. There is a period of agitation for both. Mercury either begins, or continues, along a new line of effort; or lets the matter drop.

Mercury's initial attraction to Mars is often dynamically stimulating.

Several couples had a social relationship that was continually unstable, with one or the other of them shifting into new time or energy or desire patterns that were unpredictable, and never quite reaching a blend.

There are long term marriages, family, and business relationships. In all of these there were intermittent periods of unexpected change. The businesses had fire and bankruptcy after which the working agreement of the two was reformed.

In the marriage the people moved, or had outside affairs that were disruptive, or had difficulty reconciling their sexual patterns. Where their lives would be smooth for a year or two, there would be a period of disruption at some indeterminate time.

There were two accidental deaths, once of Mercury, once of Mars. In one case Mercury was jailed for bigamy; in another Mars went into the Army. In several cases, Mercury had minor accidents that were upsetting to Mars.

The progressions are indicative of outside intervention that is disturbing to both. In one example two women friends with Mercury sesqui-square Mars were both divorced at the same time; they decided to live together.

In another example Mercury's father (who was Mars' husband) was killed by accident.

In the case of two co-workers, the business went broke, leaving them both without their pay. They commiserated with each other over a drink, and separated.

In a greater or lesser degree, the sesqui-square has an explosive effect, after which Mercury can pick up the pieces, or leave them.

Mercury Inconjunct Mars

At some time Mercury is strongly attracted to Mars in an expression of energy, trauma, stimulation, or hostility. Mercury extends his initiative to cope with the matter. In time his energy disperses into other channels but the impact of the association remains with him.

These examples show relationships that begin with the stimulation of sexual attraction, social competition, or business partnership. The enthusiasm and initiative is directed into the development of the matter.

In other examples of long term relationship, there are specific times when Mercury is drawn to Mars for a particular reason to act.

Two brothers had Mercury inconjunct Mars. They often expended their energy in common pursuits, mainly athletics and politics. Mars was injured in the war, later he had a problem with his back. At the age of 43, he was murdered. His brother Mercury, in shock and grief, extended courage and initiative in handling, with his brother's widow, the public arrangements of a funeral that caused national mourning.

A child had Mercury inconjunct the Mars of her father when he broke his leg. She energetically took the initiative of running errands and helping him in his office.

In the case of two friends, Mercury cared for Mars when he had a bleeding ulcer hemorrhage.

The progressions show a time of excitement in sexual attraction, danger, competition, or hostility.

In the most diverse example, a son with Mercury inconjunct his mother's Mars was separated from her at this time, as he was in the South Pacific during World War II. His energies were directed to bloody war battle, but he took the initiative in writing her constructive letters to keep up her courage.

One couple began a marriage, two others dissolved their marriages.

In milder examples, the two had a burst of energy for a common work or social affair.

Mercury North Parallel Mars

Mercury has a different pattern from Mars in his manner of expressing his aggressions, his energies, and his sexuality. They have had diverse experiences in these matters, with some few similarities.

They come together with spontaneous circumstantial ease, in a work environment; with social or sexual attraction; or at a time of trauma. They have a bond on a common matter of the expenditure of energy. As there is a responsive cord, Mercury will develop this as long as it fits in with his primary life pattern. As he has difficulty in changing the situation to the action he desires, his maximum expression of the planet's nature is in an area apart from Mars. In the examples of sexual diversity of Mercury N parallel Mars, one or the other may be lusty, virginal, homosexual, or promiscuous.

One marriage endured for 18 years, but it was a matter of courageous endurance. Mars was an angry, jealous and violent man, who periodically beat Mercury. When the wife's progressed Mercury repeated her natal aspect of N parallel Mars, the husband, in a fit of rage, broke her shoulder and ribs in a beating and stomping. In fear and trauma, Mercury had him jailed until she could start divorce proceedings.

Two noted politicians have the Mercury N parallel Mars. We do not have on record their personal blend or conflict, but their public association was at a time of national criticism and conflict. After their term of office, they separated.

A proper married woman, Mercury, went to work for Mars, a virile, dynamic man. She was sexually attracted to him, and found the job itself stimulating. Of course, she was impelled to express her primary energies at home.

The progressions show a period of bond when Mercury's drives are constructive; of separation when Mercury's drives are destructive. In all cases Mercury is expressing his primary energy of the matter elsewhere.

A young man with Mercury N parallel the Mars of a virginal girl, spent amorous months directed toward her seduction, while having a lusty sex life with another woman. He succeeded in the seduction, but stayed with his lover.

A woman with Mercury N parallel the Mars of her daughter cared for the dressing of her wounds after the girl had surgery on a malformed foot. At the same time the mother maintained her primary energy drives of her own passion for sculpture.

In a most harmonious example, a man had Mercury N parallel his wife's Mars for 6 years. They had a satisfactory sexual exchange, while his main vigor was directed toward his business production, with an ambitious aggression. At the end of the six years, the business affairs diminished in demand, and the home life continued on its constructive basis.

The examples hold trauma, but seldom violence. In a few cases Mercury or Mars had an accident or operation that was a shock to the other person, but did not require the other to make any effort of involvement.

In one case of a woman with Mercury N parallel the Mars of a man, both were married to other people. They found each other stimulating, but had no more than a social friendship. At the time of the progression, Mars had an operation. Mercury commiserated, but was involved in her own creative function of a first pregnancy.

The Mercury N parallel Mars is separative to the two only to the extent that Mercury must express his primary energy patterns, construction or destruction in other areas.

Mercury South Parallel Mars

Mercury has a different pattern from Mars in his manner of expressing his aggressions, his energies and his sexuality. However they have large areas of similarity in attitude or experience, that overlap.

They come together as a natural result of the circumstances of their work, sex or energy output. Due to that diversity in their basic patterns, the contact affects them in different ways. The matter implies a tension.

It is a positive, active contact, where Mercury is impelled to direct his construction, desires, or energies toward Mars.

In a marriage of 7 years, a man's Mercury was S parallel his wife's Mars. They met as a natural circumstance while he was writing his thesis for his scholastic major in music, as she was a fine pianist.

He had experienced many amorous adventures, compared to her innocence. Throughout the marriage their sex life was eminently successful, due to his skill, devotion, and tutoring. However their energy passions and ambitions became increasingly diverse, until the only familiarity they had was in bed. They developed a tension of ambivalence for several years, until they separated.

In another example, two men became friends. Mercury felt impelled to develop the friendship, and put his energies into bringing out the latent qualities of the more timid Mars. Mercury had a full and active life, in which Mars held only a certain portion of time and attention. The friendship continued on an intermittent basis of closeness and tension for years.

The progressions show a period of energetic tension as Mercury feels he must direct his action toward Mars. The contact affects the two people in different levels of importance and effort.

In several cases, a psychologist expended a constructive effort toward the rebuilding of courage of distraught patients.

There was one short sexual contact in which a man, Mercury, made a sadistic conquest of the woman, Mars, by forcing on her a drug reputed to have aphrodisiac qualities.

In several examples Mercury, both as man and woman, had intense sexual desire for Mars. While their similarities would bring them together, their diversities would hold them apart. Where they could blend on one level of bond, they could not affect a total integration. They could have social aggression, or work together, and not consummate a sexual relationship. In the examples where there was a robust sexual exchange, their energy patterns and desires would direct them into other fields of interest and activity, or competitive tension.

Because their affair in common received two different reactions, there was in some cases a covert hostility, or rejection from one to the other.

In one example, Mars committed suicide. His brother Mercury was impelled to handle his funeral and estate. The two boys had experienced from youth a highly ambivalent relationship of love, competition, and periods of rejection.

In another example, a man had Mercury S parallel the Mars of an attractive woman. They fell so passionately in love, that he left his wife to live with Mars. With satiation, she tired of his love making and turned her aggressions toward another.

The Mercury S parallel Mars indicates the need for Mercury to fulfill a constructive relationship with Mars, while maintaining his basic energy patterns in his own areas.

Mercury Split Parallel Mars

The maximum contact that Mercury has with Mars is on a matter of passion, trauma, or increased expenditure of energy in action. Both have a level of aggression or construction apart from each other, that is uninteresting or incomprehensible.

They can blend in a unity of common effort for work, social or athletic games, sex, or business cooperation, to the extent of their equality of constructive common effort. Where the two are open, positive and direct, the aspect tends to impel them together.

Where there is covert hostility, lust, force, or trauma from one or the other, or from that level in-between them, the two are forced apart.

The majority of examples hold some degree of trauma between the two, or in their total environment. There was a case where Mercury raped Mars. In two cases Mercury had a violent death, separating him from his Mars child, and Mars brother.

In one example of a man with Mercury split parallel Mars to his wife, he was a hippy king, she was 12 years his senior. Their maximum contact was in a lusty sexual exchange. His level of construction apart from her was his hand craft; her personal level of aggression was in bi-sexual adventures. They had a marriage of cooperation based on their equality of passion and sensuality.

In contrast, two women with Mercury split parallel Mars sung together in a church choir. Their personal lives were quite diverse in their enthusiasms. The maximum contact between the two women was in their jealousies and competitiveness that kept them in an armed formality of distrust.

In another example, a man had Mercury split parallel Mars to his sister. Their social and sexual aggressive patterns were quite apart from each other, but they often joined forces in a common effort. One childhood summer they built a playhouse together with rousing argument and camaraderie.

The year that Mars was 16, she broke her arm in a car accident the week before her brother left for the Army.

As adults, Mercury and Mars went into business together. They bought a hardware store that they operated with cooperation and constructive effort.

The progressions show a period of time when the contact between the two people may reach a maximum of no more than a sexual awareness that is exciting, or a common effort toward a particular job.

Love affairs were begun, others were terminated at this time.

In some cases it is a period of tension, hostility and rejection, where their lack of unity in desire clashes.

The heaviest trauma that rends the two apart can occur to Mercury or to Mars. Mercury killed himself, leaving his Mars wife. Mercury had a shock of trauma when her Mars father was beaten. Mercury fired her Mars employee for theft. Mercury's brother Mars was shot.

In one example Mercury, as an actor, played a role in which Mars was the surgeon who operated on him.

The Mercury split parallel Mars has a wide range of intensity and effect based on the extent of the trauma, passion, or effort of both persons' separate life patterns. It does have violent examples, and muted examples, that can only be evaluated fully with a careful judgement of the individual charts.

The Chart comparison on page 98 illustrates violence. The man and woman are Tony and Opal.

Tony was an Aries, with Venus conjunct Jupiter in the first house, opposition Neptune. He was a spoiled, impetuous, small time hustler, involved in various shady deals. He was slender, wiry, and handsome.

Opal had Sun, Mercury and Mars conjunct in Aquarius, Uranus in the 7th house. She was continually involved with poor choices in conduct and relationship. She was attractive, with a good figure. When she met Tony she was working as a bar girl in a gambling casino and making good money.

(Mercury sextile Mercury, natal.) They were drawn together easily, with an interest in each other and the same matters.

(Mercury semi-sextile Asc, progressed; Mercury conjunct Asc, natal.) Opal was attracted to Tony's good looks and appeal. She

Chart Comparison Illustrating Violence

Natal	Tony to Opal	Natal	Opal to Tony
☿ 6 ♈ 45 1 N.10	⚹ ☿ 3° ☌ ☽ 10° ⚹ ☉ 1° ⚹ ♂ 5° ⚼ ♃ 3° △ ♄ 1° ☍ MC 1°	☿ 9 ♒ 10 19 S.57	⚹ ☿ 3° □ ☽ 3° ∠ ♀ 1° ∠ ♃ 1° □ ♅ 7° ☌ Asc 2°
Progressed		**Progressed**	
☿ 2 ♉ 12 9 N.46	□ ♂ ∠ ♅	☿ 10 ♓ 08 8 S.15	Split P ♀ ∠ ♀ ∠ ♂ ∠ ♄ ⚺ Asc

was accustomed to shills, gamblers, and hangers-on in general, and felt an easy familiarity toward Tony from past experience as well as personal identification with his attitudes.

(Mercury sextile Sun; Mercury sextile Mars, natal.) Tony related well to Opal. He praised and flattered her, gave her credit and understanding. He helped her paint her apartment, and easily took the opportunity presented to move in with her.

(Mercury conjunct Moon, natal.) They lived together for four months. Tony had an instinctive familiarity with Opal's moods and feelings and they were highly responsive to each other.

(Mercury semi-square Venus, natal and progressed.) Opal's feelings for Tony vacillated between adoration and irritation. She couldn't understand his laziness so far as routine work was concerned, when he would plot complicated schemes by the hour.

(Mercury square Moon, natal.) Where she was concerned for his welfare and attitudes, she often withheld sympathy.

(Mercury trine Saturn, natal.) Tony easily let Opal take the financial responsibility for their living arrangements, and during the frequent times that his funds were low, he borrowed money from her.

(Mercury sesqui-square Jupiter, natal.) When Opal lost her job, Tony became surly and disagreeable.

(Mercury semi-square Jupiter, natal; semi-square Saturn, progressed.) Opal worried about the loss of income, but was more concerned about Tony's indulgence and debts.

(Mercury semi-square Mars, natal.) She was increasingly irritable toward Tony, and their sex life was intermittent, as they spent more time fighting. Opal was jealous of his attentions to other girls.

(Mercury semi-square Uranus, progressed.) Tony couldn't understand Opal's independence or why she would be so erratic in disposition.

(Mercury square Mars, progressed.) One night during a drinking bout, Tony beat up Opal. He blackened her eyes and broke her jaw.

(Mercury opposition MC, natal.) Tony went to jail when the police broke this up.

(Mercury square Uranus, natal.) This unexpected explosive crisis of incompatibility broke up Tony and Opal's affair. They did not go back together again.

(Mercury split parallel Venus, progressed.) Opal's maximum contact with Tony had been in the love they had, as well as the emotional climax of their disintegration of relationship. Had there been an equality in emotional maturity they could have sustained each other through personality conflicts and circumstantial diversities. The hard feelings, misunderstandings, and mutual amorality forced them apart.

CHAPTER IX

Mercury in Aspect to Jupiter

JUPITER RULES RELIGIOUS and personal philosophy, business and financial matters. It is the expansion of consciousness and experience that brings joy and abundance, or conceit and self-indulgence.

Mercury makes contact with Jupiter on these matters. He has an exchange of abundance or optimism with Jupiter, on matters of good will, devotion, loyalty, hope, generosity and gratitude.

The Mercury-Jupiter aspects are most pronounced in business or personal expansion, the exchange of money, goods, and benevolence.

The discordant effect of Jupiter is in poor judgment, indulgence or experience that is excessive. This leads to irresponsibility, loss of faith and great expense.

The Mercury-Jupiter parallels indicate the primary relationship of the two people within their individual life patterns of worldly or philosophical wealth; or worldly or philosophical excesses of bad judgment.

Mercury Conjunct Jupiter

Mercury has a familiarity with Jupiter's manner of expansion. From his own experience Mercury understands or identifies with Jupiter's faith and hope, religion or philosophy, wealth or prestige.

They come together to share, or experience at the same time, a matter of abundance, optimism, or expansion.

The conjunction is neutral in itself. There are examples of antipathy as well as affinity, brief contact as well as endurance.

In all the examples, Mercury gives some beneficence to Jupiter. At the least this may be a helping hand, or a service that Jupiter pays for. Other times it is loyalty, devotion or generosity. For the most part Mercury has the greater abundance of some quality or condition. This may be talent, intelligence or capability, or it may be in business, profession, prestige or wealth.

There are two examples in which both persons have Mercury conjunct the other's Jupiter. Both of these couples have prestigious position and great wealth.

In several cases Mercury and Jupiter both benefited from a common inheritance, or from a business relationship with the same person. In an example of two men with Mercury conjunct Jupiter, both were important in their profession; Mercury had received world renown. They started a business partnership with a million dollar backing from a national corporation.

In another example a man and wife both had trust funds from separate estates.

Wealth is not always measurable in financial figures. In many cases the two people have a common religion or philosophy that brings them joy, hope and benevolence.

The progressions have examples of more intimacy. In several cases the two people went into totally new experiences together that expanded their horizons and philosophy. These included love affairs and friendships.

There was a rather unlikely combination of an old Hindu gentleman and a young housewife who met at a lecture, and began a lifetime friendship. The Hindu, with Mercury conjunct the woman's Jupiter, understood and identified with her growing needs of philosophical expansion.

In one example, a woman with Mercury conjunct a man's Jupiter, paid $5000.00 to get him out of a legal bind at the time of this progression.

There is one example of two people who met in a philosophy class, had a brief clash of hostility (Mercury N parallel Mars) and separated.

In all the examples of Mercury conjunct Jupiter, both persons are in a period of growth, hope, and optimism.

Mercury Opposition Jupiter

Mercury and Jupiter expand their experience at the same time or together. Mercury has an understanding or identification with Jupiter on a matter of faith or hope, religion or philosophy, wealth or prestige.

However the full expression of their personal or business philosophy does not blend.

They can maintain a relationship of loyalty, devotion and exchange of benefit when they have alternate periods of contact and separation. When they choose a common environment their need of expansion of consciousness and experience pulls them in different directions.

In many examples these two people love and adore each other with enduring loyal faith, but their growth patterns pull them into different directions of expansion. In the long term family relationships, one or both feel that they must express their freedom or adventure, business or philosophy in a broader area than the one they have in common.

Two men with Mercury opposition Jupiter were life long friends. For six years they were in business together in which they equally made a great deal of money.

Several times Mercury withdrew from active participation in the business. He wanted to travel. He wanted to break loose for good times and new adventures. Each time he did this, he returned with a fresh outlook and bigger ideas.

Twice Jupiter withdrew from the business and personal relationship, once to open a new branch office that he managed for eight months, and once to marry and go on an extended honeymoon.

When the business was sold, the two men parted with relief and reluctance. The faith and trust they had in each other had been shaken a few times, but it endured in solid loyalty. Both were content to go on to their personal areas of expansion, but they continued to see each other periodically with joyful warmth.

In three examples (out of 48) the two were separated by the death of a Mercury father, a Jupiter father, and a Jupiter sister.

The two fathers left a sizeable inheritance to the young child with whom they had the aspect.

In all but a very few examples there is some financial, business, or professional exchange.

Mercury was an entertainer, whose girl Jupiter watched him from backstage. They had a devoted romance for a year.

Mercury rented Jupiter's house.

Mercury was an actor who paid 10% of his earnings to Jupiter, his agent.

Mercury and Jupiter were in the same business. They had a brief love affair that was the first extra-marital experience for both.

Mercury was a dentist, Jupiter was his patient.

The progressions indicate a peak period of familiarity and separation.

A man and wife had a divorce during this period that was costly to both. They retained a loyal respect for each other, and continued to work in the business they owned in partnership.

Two men who had been friends for years bought a house. Their personal philosophies clashed at close range. Mercury felt that Jupiter was self-indulgent; there was a breakdown in trust and they separated. After a cooling off period they resumed the friendship.

The planet Jupiter is the greater benefic. Even in the separation of viewpoint or experience that often occurs with the opposition aspect, there is goodwill between Mercury and Jupiter.

Mercury Square Jupiter

Mercury is strongly drawn to have trust, hope and faith in Jupiter. He has such an overabundance of optimism that it colors his judgment. He may expect too much of Jupiter in understanding and accomplishment.

When Mercury has devotion, loyalty and gratitude, he expands his consciousness and experience with Jupiter.

When Mercury has conceit or self-indulgence, he strains Jupiter's good will to the breaking point.

In many of these examples Mercury has such a profusion of devotion to Jupiter that he feels Jupiter can do no wrong. When the relationship is based on a philosophy of honor and trust there is great mutual exchange of benefit. When Mercury overrates Jupiter's capacity, he may demand more than Jupiter can handle.

There is at some time or in some area a conflict in their philosophy of the guiding principles of life.

In one example of two women, Mercury was a plain, heavy set woman, who worshipped and adored Jupiter for her poise, glamor, and great buoyant manner of doing things in a grand style. When they met, (with Mercury trine Mercury) they had an easy and immediate communication. They found they had been born and raised in the same town and knew many of the same people.

Jupiter was in financial straits and Mercury offered her a home. They lived together for three years. Though Mercury paid the rent, she felt more than compensated. Jupiter helped her to diet, and taught her how to dress and carry herself in a public situation. They shared much of their social life, and had a firm faith in the church they attended together.

An area of conflict developed in their financial situation. Mercury began to present Jupiter with various bills, including those of personal expense. Jupiter paid them, but with a growing indignation. Their good will for each other became strained, until Jupiter moved out.

In many of the various examples there is some exchange of money or goods. Mercury gave Jupiter an allowance. Mercury loaned Jupiter his car. Mercury paid Jupiter a wage. Mercury was expensively entertained by Jupiter. Mercury paid a bill to Jupiter. Mercury borrowed money from Jupiter and didn't pay it back.

The progressions indicate a critical period of adjustment in expansion. When Mercury gives his faith, loyalty, good will, generosity and optimism to Jupiter he wins a devoted ally.

When Mercury is demanding or self indulgent there is a breakdown of trust and faith that puts a serious strain on the association.

As an example, two men friends had Mercury square Jupiter. Jupiter, at this time, was promoting a new sales technique, and he enthusiastically spent weeks telling Mercury all about it. Mercury was an important man, busy expanding his own affairs. He said privately, "Jack is such a great guy I know he'll put that program across, but if he tells me one more word about it, I'll throw him out of here myself."

But Mercury listened in loyal patience and encouragement. When Jupiter's sales program went over successfully, Mercury was able to adapt it to his own company's needs in a way that increased his profit.

In other various examples Mercury gave Jupiter an inheritance, his wages, expensive gifts, and a loan of money. For the most part these were received gratefully. Mercury never lost from his giving. When Mercury gave his faith and open good will to Jupiter, he cemented a loyal friendship.

In the cases where Mercury asserted his over-demand, this relationship strained or broke.

Mercury Trine Jupiter

Mercury has a spontaneous good will for Jupiter. Circumstances and environment are such that they fall into a natural contact under the most favorable circumstances. As the opportunity for association comes so easily, Mercury is not always appreciative.

Where Mercury is honorable, loyal and encouraging, he receives from Jupiter a full benevolence. Where Mercury takes indulgent advantage of Jupiter he loses the benefit, often at personal cost.

In many of these examples, Mercury gave freely to Jupiter of whatever he had to give; money, time, attention, praise, good will and appreciation. This was easy for Mercury to do, as circumstances placed them together favorably. In return, Mercury received benefit many times over, in whatever Jupiter had to give. In these examples the two people had an exchange of gratitude and devotion.

In the example of the Japanese student who lived with the American woman (refer to Mercury trine Moon) she had Mercury trine his Jupiter. Jupiter was in a foreign country, with a language barrier that increased the difficulty of his class work. He was separated from his home and family, and from everything dear and familiar to him. Mercury had a large buoyant family home; it was natural and easy for her to extend him good will. With their exchange of consideration (Moon) and loyalty, they built a devotion to each other.

The summer after Jupiter had returned home to Japan, he invited Mercury to visit his home. He not only took care of all transportation and accommodations, but he extended himself in making the two weeks in Japan a joy and a wonderland of enchanting experience for Mercury. She will always remember that trip with gratitude and pleasure.

In some examples, it was difficult to tell who received the greater benefit, as the exchange was not usually in the same coin. One person often expanded the experience of the other with new adventures or philosophical concepts. The other person often gave a gift, or service, or benevolence.

As the trine does not imply a great deal of effort, many of the examples held a trace of indulgence where Mercury accepted Jupiter benevolently for as long as the contact was personally satisfying. When Mercury gave money but not loyalty, there was a loss.

The progressions show this more graphically. There were several brief love affairs that Mercury began as a matter of adventure or casual good will. In some cases Mercury allowed Jupiter the largess of his bounty as long as it cost him no effort. In other examples there was an association rich in exchange of joy and abundance.

It may definitely be said of the Mercury trine Jupiter that when the exchange was that of loyalty and devotion, the benefit was greatest for both.

Mercury Sextile Jupiter

Mercury has the inclination and the chance to give Jupiter hope, encouragement, loyalty and benefit. At a certain time, the two are drawn together easily for the opportunity of expanding a good will relationship. If Mercury remains indifferent, the chance is lost.

All these examples show a definite time when Mercury has such good will for Jupiter that he gives generously of whatever he has; money, goods, approval, attention, devotion and praise. That time may be the week that they meet, or over the years that they know each other. This forms a basis for business contacts, friendship, love, and marriage that often was radiant.

Mercury gave, in the various cases, loyalty, devotion, work, gifts, entertainment, moral support, her home, her virginity, a job, his car, money, clothing or adventure.

Other than the basic philosophy of each persons' system for guiding life, there were very few examples of a specific philosophical exchange.

In an example of indifference, a young woman had Mercury sextile the Jupiter of a wealthy man. She had an attraction of good will toward him; they began dating. When he bragged of his indulgent romances with others, Mercury's interest waned, and she gave him no further encouragement.

In another romantic loss, the woman, Mercury, decided to play a coolly sophisticated game instead of showing Jupiter how much she admired him. She lost the chance, as he chose a warmer girl.

The progressions show the specific time of opportunity. In one case Mercury loaned Jupiter $8000.00 for a business venture.

In another example, a woman with Mercury sextile her husband divorced him at this time (Mercury opposition Mercury) but continued to give him hope and encouragement to ease the break. She visited his home weekly to share his bed, until they both had the emotional security to move into separate new worlds of expansion.

Excellent friendships were formed at this time.

One couple were at a point of crisis in their marriage, (with Mercury semi-square Mercury, and Mercury sesqui-square

Venus) but with the Mercury sextile Jupiter, maintained the loyalty and faith in each other to continue past the difficulties.

Mercury Semi-sextile Jupiter

Mercury has a vivid spark of good will for Jupiter. With cooperation they can expand a pleasurable contact into one of loyal devotion.

When other aspects add substance and endurance, the Mercury semi-sextile Jupiter can be the trigger to joyful enduring good will. Each act of generosity received gratitude in a mutual exchange.

Some of the examples have less duration or depth. There are short term affairs in love, friendship, or business, where there is some exchange of good will.

Mercury makes the primary effort. Jupiter's response will depend on his chart and attitude.

The progressions refer to a specific time when Mercury makes an effort to extend a benevolence to Jupiter.

Mercury gives his faith and loyalty, even without response. With cooperation, both persons expand their experience beneficially.

Mercury Semi-square Jupiter

Mercury has intermittent periods of good will and loyalty to Jupiter. As Mercury doesn't fully understand Jupiter's philosophy of life, in anxiety he may become critical. Mercury can handle these slight changes with firm patience, as conditions either right themselves or fade in importance.

In an example of a long term marriage, the wife's Mercury was semi-square her husband's Jupiter. Though she was devoted to him, she could not understand what system for guiding life impelled him to certain decisions. When Jupiter was over-expansive in the purchase of home equipment, Mercury was critical. When he applied for a job transfer, Mercury's worry and anxiety caused her to be irritable. The matter always worked out more smoothly when Mercury held her peace.

Other examples are of friendships and family relationships that followed this same general pattern.

The progressions are more often indicative of intermittent contact. In a long term relationship, the two may have some friction over finances.

In all the examples of Mercury semi-square Jupiter, Mercury feels that Jupiter uses poor judgment or indulgence in some area.

Mercury Sesqui-square Jupiter

Mercury has a sudden situation, or recurrent intervals of disruption with Jupiter in their patterns of growing consciousness and experience. This tends to break up present conditions. Mercury can decide to expand the contact under the new conditions presented, or let the matter drop.

In many of the examples, Mercury or Jupiter is having new experiences in his environment that cause agitation for them both.

In a long term marriage that was loyal and devoted, Mercury had a business failure that caused a financial pressure for a number of years. The standard of living that he and his Jupiter wife were accustomed to was broken up, but their philosophy of hope and faith in each other was unbroken.

In another example, Mercury's untimely death broke up his contact with his small son Jupiter. The boy received a fortune in inheritance.

The disruption may be directly between Mercury and Jupiter. There may be an attraction of good will between them, when circumstances are adverse and disruptive.

In one such case, a man and woman were attracted to each other; both were single and eligible. They were friends for many years, but whenever Mercury decided to develop the relationship into a different pattern, a disruptive situation would interfere. Mercury would be placed on a split shift at work, or Jupiter would be moving or going into another romance.

Mercury finally dropped the matter entirely as being too agitating.

The man who tried to kill his wife (refer to Mercury N parallel Mars, Mercury square Mars) had Mercury sesqui-square her Jupiter. He felt that she was faithless and self indulgent in wanting to go out bowling and in spending money on clothes. When she joined the church they had a disruptive period of agitation.

The progressions show an agitating time. In long term relationships, one of them may be having new experiences that preclude the other.

The Mercury sesqui-square Jupiter shows profuse changes in unpredictable areas that require readjustment if the two are to continue a contact.

Mercury Inconjunct Jupiter

Mercury is strongly attracted to Jupiter with a profusion of good will and loyalty. He takes interest and initiative in expanding their relationship. In time, Mercury's interest disperses into other areas but he keeps a mellow regard for Jupiter.

In most of these examples Mercury has a wide open devotion and faith in Jupiter. The two usually have a similar philosophy or business.

Though Mercury's expansion of experience takes him into other areas he will maintain a friendship, marriage, or business contact with Jupiter of mellow good will. In a long term relationship Mercury's faith is periodically renewed.

The progressions show a climax of profusion in Mercury's loyalty, devotion, and generosity to Jupiter.

There are two deaths in the Mercury inconjunct Jupiter examples. In the case of Jupiter's death, he left Mercury a fortune, as well as a memory that was devoted.

In the other case Mercury was a suicide, who left a profuse grief to his old friend Jupiter.

The presence of Jupiter aspects at the time of death are indicative of inheritance in the appropriate cases. Philosophically, death can be considered as the supreme experience of expanded consciousness.

Mercury North Parallel Jupiter

Mercury has a different pattern from Jupiter in his expansion of consciousness and experience. They have had diverse backgrounds in their philosophy, or in their financial or business environment, with some few similarities.

They come together with spontaneous circumstantial ease, for a business or personal expansion. They have a bond on this matter in common, that hits a responsive cord in both persons. Mercury will develop this to the extent that it fits in with his primary life pattern. Mercury is impelled to his maximum expression of the matters in an area apart from Jupiter.

In many of these examples there is an exchange of benefit. The nature of the benefit is indeterminate without a full knowledge of what it is that each person has to give; nor is it always an equal exchange.

In one example, Mercury gave her father the joy of her buoyant youthful exuberance, with hope and good cheer through his failing years. There was a response of loyalty and devotion. Upon his death he left her an inheritance of property and money.

Two young men with Mercury N parallel Jupiter shared an expansion of consciousness and experience with drugs over the two year period that they lived together. At the same time, Mercury was expressing his maximum effort to expand his business contacts, apart from Jupiter.

In all of the examples, whatever responsive bond the two people had, Mercury expressed his larger areas of business, philosophy, or experience apart from Jupiter. He cannot seem to change their situation to as full a benefit as he would like.

The progressions show a time when there is a bond of good will and benevolence. Mercury gives freely to Jupiter of his time, money, attention, praise and loyalty. But the bond cannot be complete: Mercury must still maintain the larger expression of his benevolence in an area apart from Jupiter.

In a devoted and adoring friendship between a woman and a man, the woman Mercury maintained the solidarity of her first loyalty to her husband and home.

Two men had an inseparable friendship, in which they had jovial good times together and worked in a similar business. Mercury had an area of religious devotion that he held sacred and did not share with Jupiter.

In one case, a man with Mercury N parallel the Jupiter of his girl friend spent over $1000 on entertainment alone during their four month affair. Jupiter did give him new adventures and good times for his investment. Mercury eventually had the need to express his own scope in his neglected business, and his time and attention left the girl.

The Mercury N parallel Jupiter is separative only to the extent, and in that area, that Mercury must express his primary patterns of expansion, philosophy, or business apart from Jupiter.

Mercury South Parallel Jupiter

Mercury has a different pattern from Jupiter in his expansion of consciousness and experience. They have large areas of similarity that overlap in their philosophy, or business experiences.

They come together as a natural result of circumstances and environment. They have a great deal in common. Mercury is impelled to expand their relationship in terms of hope and joy, religion and philosophy, business or finances. These matters affect them both.

Mercury has a tension in his good will because of their diversities.

In many of the examples, Mercury is impelled toward Jupiter in radiant joy and devotion, hope and faith. They have similarities in philosophy, or in business or financial matters. The contact expands the consciousness and experience of both persons. They have an exchange of abundance and optimism.

The tension in the contact lies in the poor judgment or self indulgence of one or the other. When irresponsibility enters the relationship, the exchange becomes reluctant and costly.

Because of their diversities, Mercury must maintain areas of expansion apart from Jupiter.

In a youthful marriage, the girl's Mercury was S parallel the young man's Jupiter. They were impelled together in adoration, joy, hope for the future and faith in each other. Neither took a great deal of responsibility, though their patterns of indulgence were diverse. A tension developed in these areas. Mercury began to feel a need of expanding her experiences in areas apart from Jupiter, and they separated.

A woman and her son both had Mercury S parallel Jupiter to each other. The woman was wealthy, and she indulged the son. Whenever he began another costly expansive business or adventure, mother paid the bills. She did this reluctantly, and often there was an ambivalent tension between them, but the pattern was lifetime. They adored each other. The multiple S parallels created a compelling tension of response.

The progressions show a period when Mercury finds Jupiter irresistible. Even if reluctant or ambivalent, Mercury is impelled by a tension of attraction to expand their contact.

A man with Mercury S parallel his wife's Jupiter invested $10,000.00 into a business she wanted. She lost the money in five months. Mercury's own business was in a critical decline, and Jupiter had to take a job to pay the rent. The financial tension reflected on their personal relationship. These two people loved each other, but both felt that the other had poor judgment and self indulgence.

Another man with Mercury S parallel his wife's Jupiter was divorced by her at this time. Their patterns of expansion were in a diverse period; both went into new experiences and financial indulgence. Mercury was reluctant to pay alimony, but in this case was impelled to do so by law.

The Mercury S parallel Jupiter shows expanding life patterns. There is often the greatest happiness, but it is also costly in some way. The aspect is unifying to the extent and in the areas where Mercury must fulfill a meaningful relationship with Jupiter while maintaining his primary life pattern of hope and faith, religion or philosophy, business or finances.

Mercury Split Parallel Jupiter

Mercury's maximum contact with Jupiter is on a matter of the expansion of consciousness or experience; religion or philosophy; or business and finances.

Each has a level of environment and experience in these matters that is apart from the other. These areas are uninteresting or incomprehensible to the other.

They can blend in a cohesive unity in ratio to their equality of faith and trust, honor, candor, and cooperation. When their separate lives hold hidden factors of irresponsibility of self indulgence it tends to force them apart. When their areas of personal expansion hold incommensurate factors, they tend to be forced apart.

The examples show people in broad areas of expansion in their life patterns of experience. When they unify, the harmony and benefit show intense loyalty and radiant devotion. In the business examples, unity can lead to fortunes.

When there is covert or open indulgence or irresponsibility, the contact ranges from profuse antipathy to overwhelming expense.

A man who molested a child had Mercury split parallel her Jupiter. His indulgence was costly to her in a loss of faith and trust that marked her life. The cost to him was of his home and personal liberty, as he was jailed and divorced by his wife.

In one example a man and woman both had Mercury split parallel the Jupiter of the other. When they met, they were overwhelmingly impelled together with a profusion of emotional attraction. They felt that they had to be together at any and all cost, that all happiness could be theirs together. They were both married to other people; they had a combined total of seven children.

The man and woman ran away to Mexico, to get quick divorces. The relationship, built on poor judgment and irresponsibility, also lay host to secret indulgences in their separate areas apart from each other. The man had a history of debts and a penchant for writing bad checks. The woman outfitted their house with new and expensive furniture that she charged.

They were ecstatically happy and deeply in debt when the law intervened. The man was arrested for larceny and bigamy. He was jailed, and later released under psychiatric care.

The whole affair was costly to the woman in peace of mind, profuse emotional strain, loss of faith and loss of her home and children. When the man began to repeat his pattern of writing bad checks, they separated.

In contrast, there is an example of two men with Mercury split parallel Jupiter. They were loyal friends for a lifetime. Both expanded their separate lives to a maximum of success in their own fields, to become wealthy and well known for prestigious accomplishment.

Both of these men had a basic honor and trust in each other. Early in their careers, at a time when Mercury had not enough money to pay his rent, Jupiter handed him several hundred dollars to get a start. He also gave him a job. Mercury did this job so well, that he was able to repay the loan and go on to his own business within two years.

He was soon clearing a quarter million dollars a year. The two men were able to handle cross contracts that benefited each other. Perhaps of even more value was the esteem of devotion and loyalty that they had for each other.

The progressions hold a wide range of examples of specific times of expanded experience. There may be as pure an exchange as that of loyalty and faith. A mother had Mercury split parallel Jupiter to her child at the time when they moved to a more expensive home, and the child began school.

Two men with Mercury split parallel Jupiter were in the same business; both were given increases in wage and more prestigious positions.

A man with Mercury split parallel Jupiter to a woman began a love affair with her in an exchange of faith and trust that exceeded any previous experience of either.

In the examples when there were hidden factors in their separate lives, the two people tended to either drift apart or to be forced apart. A wealthy man who was murdered had Mercury split parallel Jupiter to both his brother and his son.

A man with Mercury split parallel Jupiter to his employer was guilty of petty thievery and was fired.

In some cases the expanding areas of the two people were so foreign to each other, it created an unbridgeable gap.

When this relationship reaches a maximum of unity in sharing a cooperation of philosophy and good will, it cannot be excelled for happiness and benefit.

The Chart Comparison on page 118 illustrates a casual advantage and rejection in a short term relationship. There is no Mercury to Mercury aspect, natally or by progression.

Shirley had been born to a wealthy family, and educated in private schools. As an independent Sagittarian, with the strong will of Pluto in the first house, she left the familial security to go her own way, and took a job and apartment in another city.

Will was an ambitious aggressive Capricorn with Mars in the first house. By the age of 28 he had a firm position in a sales firm that paid him well. By shrewd independent investment he was accumulating a small fortune.

(Mercury sextile Mercury, natal.) The opportunity of their meeting occurred when Shirley was commissioned by her employer to carry several contracts into the sales firm to deliver personally to Will.

(Mercury semi-sextile Mars, natal and progressed.) They both had a vivid spark of sexual awareness of each other. Will asked Shirley if he could call her.

(Mercury square Mercury, progressed.) Shirley was engaged to marry another man. She reacted to Will with a wary formality and told him she was not free to receive calls.

(Mercury sextile Neptune, natal.) The meeting was disturbing to Shirley. She found herself dreaming of Will, and picturing him in romantic fantasy.

(Mercury trine Uranus, natal.) The matter may have ended then, if a friend of Wills had not insisted one night that Will accompany him to a lecture on Eastern philosophy.

This was a study that held Shirley's sincere interest and she was there without her fiance. A small group went out for coffee

Chart Comparison Without Mercury to Mercury Aspect

Natal	Shirley to Will	Natal	Will to Shirley
☿ 15 ♍ 01 ℞ 17 S.45	⚹ ☿ 1° ⚹ ○ 3° ☌ ♃ 2° ⚹ ♆ 1°	☿ 13 ♉ 40 24 S.38	⚹ ☿ 1° ⚹ ♀ 6° ⚺ ♂ 1° △ ♅ 7°
Progressed 1951 ☿ 10 ♍ 28 24 S.42	S.P ☿ ⚺ ♂	Progressed 1951 ☿ 14 ♒ 19 ℞ 13 S.45	S.P ♀ Split P MC □ ☿
Natal ☿	S.P ☉ r S.P ♃ r □ ☿ r	Natal ☿	S.P ☿ r

after the lecture and Will questioned Shirley with intelligent respect.

The lecture series ran for three more weeks. Will went to each one as he found the subject provocative, as well as being magnetically drawn to Shirley.

(Mercury sextile Venus, natal.) Opportunity again presented itself when Shirley's car had a flat tire on the last night of the lecture series, and she had no spare tire. Will drove her home. They talked, and kissed.

(Mercury S parallel Mercury, progressed.) Shirley and Will were impelled together in a close, intense communication and affinity, as they continued to see each other. Shirley broke her other engagement.

(Mercury S parallel Venus, progressed.) Will hadn't planned to marry. He began to build a tension of ambivalence, in resisting the love he felt for Shirley. He dated other girls, but found them pallid. His work suffered because of his emotional involvement.

Shirley refused an affair, and Will accused her of setting a price on her love. As the aspect waned, Will's resistance to happiness grew less, and he adored Shirley.

(Mercury sextile Sun, natal.) Shirley could understand and appreciate Will's goals and character. They had an exchange of deep respect and credit.

(Mercury conjunct Jupiter, natal.) From her own background, Shirley could identify with Will's areas of expanding experience, both in business and in his personal philosophy.

(Mercury S parallel Jupiter, progressed.) Because Shirley had already known the experience of wealth, she was indifferent to this for herself, and was more concerned with her development of spiritual awareness. She could recognize Will's drive to material success without being in agreement with its value.

The tension of attraction between them was radiant. She walked through her days with the light of happiness shining from her face.

(Mercury S parallel Sun). They planned to marry. Shirley was impelled to relate to Will with the deep bond based on similar integral patterns. Because of her independence and the satisfaction she had in her own work, she insisted that she continue her job.

(Mercury split parallel MC, progressed.) Shirley's family sponsored a social wedding that was formal and complex.

Sixteen years have elapsed since that wedding. There have been joys and sorrows, problems and successes that have repeated the natal aspects in other experiences. Shirley and Will have kept their exchange of respect and understanding and their love for each other.

CHAPTER X

Mercury in Aspect to Saturn

SATURN RULES WORK AND RESPONSIBILITY, for the goal of emotional and temporal security and safety.

Mercury makes contact with Saturn on these matters. There is an exchange of security and stability, system and endurance, in a mature and responsible contact.

The Mercury-Saturn aspect often implies a problem or grief, hard work or loneliness, lack or loss, depression or discouragement. The discipline of Saturn can be self discipline or externally imposed restriction.

The most discordant effect is of selfishness, greed, underhanded deceit, dishonor, cold loathing, or tragedy.

The Mercury-Saturn parallels denote a serious relationship between two people within their individual life patterns of maturity and responsibility.

Mercury Conjunct Saturn

Mercury has a familiarity with the manner in which Saturn expresses responsibility. He understands or identifies from his own experience with the problem or grief, hard work or loneliness, restriction or depression of Saturn.

They come together to share, or experience at the same time a mature responsibility or a severe problem.

As the conjunction is neutral, Mercury can acknowledge Saturn impartially.

These examples include long term marriages, family contacts and friendships. Even when they are not directly business contacts, often there is a matter of business involved.

Saturn in all cases was hard working, or had some grief or problem, to which Mercury would give some acknowledgement or responsibility. Mercury too, knew well the heavy load.

The exchange could be as impartial as in the case of a dentist with Mercury conjunct his patient's Saturn.

In one case, a priest had Mercury conjunct the Saturn of a lonely, hard working woman. He could understand the discipline of her life from his own experience. He gave her the comfort of traditional stability.

In an example of a marriage, a woman had Mercury conjunct her husband's Saturn. They had many years of hard work and security together, as well as periods of severe problems, in their business, health, and personal discredit.

Another marriage was a model of mature responsibility. The warmth and love in it came from other aspects, as the planet Saturn is at no time warm. These two people were poor and hard working. They gave each other security and stability.

Other marriages and relationships died from selfishness and lack of joy.

The progressions are indicative of a time of problem or effort. Two lovers had Mercury conjunct Saturn when Saturn shot himself. They were breaking up, and Saturn was in a depression of grief and rejection.

A mother had Mercury conjunct her daughter's Saturn when the daughter's husband drowned.

A child had Mercury conjunct his father's Saturn when his parents separated. The boy identified with the father in a feeling of loss.

At the least, the Mercury conjunction Saturn implies a cool exchange. In the Mercury-Saturn aspects, the people may work together. They may well have a mature respect for each other. They do not have play, joy, or pleasure in the Saturn exchange.

Mercury Opposition Saturn

Mercury has in common with Saturn a similar work or responsibility, grief or problem. Where they reach an understanding on this matter, the full expression of their viewpoint or experience in work, self-discipline or restriction does not blend.

This is the greater malefic in the most discordant aspect. It implies an inexorable struggle of selfishness and problems. The two people may have a steady contact in a work environment with a responsible endurance of duty; or they have at some time a problem or grief between them.

Apparantly many people need this discipline, for we have, surprisingly, long term contacts in marriage, family and business. In these there is at some time a dishonor, discredit, or loss.

The severest examples have underhanded deceit, cold loathing, or tragedy.

One woman married twice; both times she had Mercury opposition the man's Saturn. She lived for 20 years with her first husband, a cold, reserved older man who gave her financial and social security. It was a stable, dutiful marriage, if a lonely one for both of them.

After his death, the woman remarried. This time she took the Saturnine role of being older, hard working and responsible. She and her husband went into a series of business ventures that made gain with effort and discipline. Their 10 year marriage held certain joys, and many personal problems and anxieties. Under the stress of a business failure and loss, they separated.

In some examples, the two people had no commensurate responsibility on which to base a relationship. They met and parted with distaste or contempt.

In other cases, the two people loved each other, but the life of one or the other was marked with tragedy. A woman had Mercury opposition the Saturn of a dear friend who went blind.

In various examples, Mercury was assassinated; Mercury was imprisoned for his own misconduct; Mercury lost his business; Saturn died at an early age; Saturn was in an auto accident where he was badly broken up; Saturn was homeless and alone; Saturn had a series of accidents and illnesses.

A son with Mercury opposition his mother's Saturn cared for her financially from the age of 17 when his father died, to the time of her death 30 years later. The relationship was respectful, but cold.

During the progressions, the two people can meet very well on a matter of business. This still may be costly in some way, usually to Mercury.

One woman, Mercury, invested $10,000.00 into a business that Saturn began, and lost it in 6 months.

Several love affairs had disastrous results. In two cases Mercury was selfishly used and abandoned by Saturn. In another case a Mercury man lost his reputation when his girl friend Saturn had an abortion.

The least harm in the Mercury opposition Saturn examples was in the cases where both persons took a hard working, responsible attitude.

Mercury Square Saturn

Mercury is strongly drawn to take on work or responsibility for Saturn or develop a working relationship. There is either a circumstantial obstacle or an incompatibility in their areas of responsibility.

A mature and unselfish Mercury will help Saturn through grief or loss, poverty or problem, job or work.

A self-centered, cautious Mercury will coldly abandon Saturn to his own burden.

In some of the examples the obstacles are such that Mercury cannot give Saturn the type of help he needs. In an example where Saturn had a failing business, Mercury didn't have the money to give him financial assistance, and had to go to work somewhere else to support his own family.

In other cases, the two were separated by distance at the time when Saturn had a need; or Mercury had other commitments that had to come first.

The various problems that Saturn had were in health, injury, loneliness, harsh discipline, over-work, poverty, divorce, death in the family, jail, or psychological difficulties.

In an example of two men with Mercury square Saturn, Mercury went to work for Saturn, to handle his problems of cataloguing a series of records into a methodical system. Mercury was interrupted in his work by an accident in which he broke his leg.

A mother had Mercury square the Saturn of both her sons. They were not "problem children" in any sense of the phrase.

One of the sons had several accidents resulting in broken bones during his childhood years, and a general physical weakness. The other son was a serious, responsible boy. The mother worried about his lack of frivolity or fun during his adolescence.

The Mercury square Saturn is primarily a responsible, responsive exchange. Most of the examples are of long term relationships, in family, friends, marriages, and business contacts.

There may be no more than a period of difficulty that can be overcome with effort.

In an irresponsible example, two men were business partners. Mercury managed the business, Saturn put up the capital. Mercury worked as hard as he was able, but he did not have the system, stability or maturity to handle the management. When the business was totally lost to creditors, Mercury's attitude was, "Well, too bad," and he went on to other fields, leaving Saturn with the heavy financial loss.

The progressions show a particular time when Saturn was hard working, mature and responsible, had a grief or loss, was poor or in debt, or was ailing, physically or mentally.

Mercury offered help and assistance to the extent that he could; his stability of friendship, his business efforts, his time and attention. Often their conflicting responsibilities kept them apart.

Mercury Trine Saturn

Mercury has a spontaneous and natural contact with Saturn in which he can easily handle any difficulty that arises. Mercury is seldom appreciative of the depth or extent of Saturn's responsible capabilities, or of Saturn's discipline or restriction.

A mature Mercury will draw upon Saturn's depth to develop an exchange of mutual benefit. A self-centered Mercury makes an indifferent use of Saturn or takes advantage of him.

The majority of examples show two people who could so easily give each other security and stability. Mercury seldom makes any effort beyond the natural exchange of circumstances.

A man had Mercury trine the Saturn of a youth who was a family friend. The man said, "Sure, Bobby's a good, steady boy. He's heavily disciplined at home, but he'll make out all right." When Bobby ran away from home, Mercury wondered if he could have helped more than he had.

A man with Mercury trine the Saturn of his employee easily gave the younger man a variety of responsibilities. However, when an opening came for a position requiring methodical system and capacity of management Mercury overlooked Saturn in favor of another. Saturn could have handled the job more efficiently than the other man. He proved this when he took a similar position with a competitive company.

In several examples of marriage, friendship, and business exchange, Mercury did credit Saturn with stability, maturity, and capacity. In these cases the two worked together with emotional and temporal benefit, that built a secure relationship.

In the progressions, Mercury often leaned on Saturn for security, rather than develop an exchange.

In other cases, Mercury underrated Saturn, when Saturn was exactly the person who could have helped best.

When the two put in a common responsibility toward their relationship, it was easily satisfactory.

Mercury Sextile Saturn

Mercury has the ability to develop a secure and responsible contact with Saturn. There is a certain time when the way is clear for them to build a mature exchange. Mercury may offer work, stability or security at a time when Saturn has difficulties. A self-centered Mercury loses this opportunity by default, or by taking advantage of Saturn.

This is an advantageous aspect for both persons. In some cases they build a mature respect for each other that is enduring.

In other examples, Mercury can't be bothered. He does what is required and no more. The contact may be lost entirely or become no more than perfunctory.

In the various cases, Saturn's difficulty may be in business efforts, over-work, loneliness or poverty, death in the family, depression or discouragement. These are not long term burdens, but rather incidents that occur during the time of Mercury and Saturn's association.

In a few examples of total irresponsibility, Mercury took selfish advantage of Saturn, financially, sexually, or with deceit. They parted in cold loathing of each other.

The progressions show a period of time when the two people may be very close as Mercury responds with comfort or assistance to Saturn's difficulties. In various cases, Saturn was divorced, had legal problems, poor health, overwork, discipline or restriction.

Mercury did not always choose to take the responsible step. In these cases the contact remained formal, or was lost.

Mercury Semi-sextile Saturn

Mercury has a vivid spark of interest in whatever difficulty Saturn may have. If both persons put in a mature effort, they can develop a responsible association.

Neither one is without problems of some nature, but they could help each other if they both make the effort.

A woman with Mercury semi-sextile the Saturn of her employer was quite willing to work with him during a business crisis on partial wages. When he didn't pay her anything, her own bills accumulated and she had to quit.

A man with Mercury semi-sextile the Saturn of his friend would have helped Saturn with his personal problems, with comfort and advice, but could not interfere without being asked.

The progressions often show a single problematical incident or short period. Mercury is willing to help but not always able to do so. Saturn, in the various cases, was psychologically disturbed and depressed, injured, ill or weak, or in a new area looking for a secure residence.

Several marriages began at this time, with the two people sharing the constructive problems of house purchase or new jobs, or adjustment to responsibility.

Mercury Semi-square Saturn

Mercury has intermittent worry and anxiety about Saturn's problems or responsibilities. He doesn't fully understand the nature of the difficulty and may be critical or irritated.

Mercury can handle the slight changes in their contact with firm patience as conditions will right themselves, or lose importance.

In most of the examples Mercury has a solicitous concern for Saturn and takes a responsible attitude.

A man with Mercury semi-square the Saturn of a friend realized that he was troubled and overworked. Mercury didn't understand the depth of Saturn's insecurity and depression. There was little that Mercury could do but offer his own stability of friendship. The period passed and Saturn overcame his debts and improved his attitude.

A girl had Mercury semi-square the Saturn of two of the boys with whom she associated in the same school and social group. She worried and fussed at them to get jobs and avoid trouble with the authorities. Their difficulties straightened out with time, and her interests went on to other areas.

The progressions also show examples where there is not much Mercury can do other than express his concern.

In one case Mercury did intermittent short jobs for Saturn, that caused him some anxiety as he wasn't sure that Saturn approved of his work.

In several examples Mercury had a flare up of worry and criticism of Saturn, when he did not understand Saturn's attitude toward responsibility.

A Mercury child had a period of insecurity that made a deep impression, when her Saturn father died.

Mercury Sesqui-square Saturn

Mercury has a sudden situation, or recurrent intervals of disruption in the work and responsibility he has with Saturn. This breaks up their present conditions. Mercury can readjust their contact on a more mature level, or let it drop.

The sudden situation is usually a matter in which the areas of responsibility of one or the other changes. This causes a period of insecurity or restriction.

In the case of two men who were friends, Saturn took on a new business. Mercury quit his job to work with Saturn for less money, but a percentage of profit. The business failed, causing a loss that was agitating to both men. Mercury took a stable and mature attitude toward Saturn that saved the friendship, but they did not work together again.

There was one marriage begun with a Mercury sesqui-square Saturn, but the aspect is not favorable to romance. The work and perseverance over agitative problems is too disruptive.

The progressions show periods of grief, loss or hard work. There is an example of a floundering marriage that survived, and a business partnership that died.

A child had Mercury sesqui-square his mother's Saturn at the time that she and his father separated. The breakup of his home conditions threatened his security and caused him agitation.

Mercury Inconjunct Saturn

Mercury is attracted to a work or responsibility with Saturn. He takes an initial interest that disperses quickly. Mercury's efforts either go into other areas of his own circumstances and interests; or he has a disappointment in Saturn.

Other aspects must be considered for substance. In many examples, Mercury has no more than the urge to responsible contact with Saturn that is not carried through into action.

A man with Mercury inconjunct his brother's Saturn acknowledged his hard work and poverty, but had his own responsibilities to care for.

In stronger examples the contact was more of a problem. In three cases of business partnership there was failure and loss. In one case Mercury seduced and abandoned Saturn. A Mercury child lost her Saturn father in death.

In the progressions, Mercury often made a single feeble effort to work with Saturn and abandoned the whole idea as a bother.

In other cases Mercury became involved in a problem with Saturn that was discrediting to both.

A wife with Mercury inconjunct her husband's Saturn maintained their security during the time when he had a heart attack.

Another couple had a period of debt.

Mercury North Parallel Saturn

Mercury has a different life pattern from Saturn in work and responsibility, discipline and restriction. Their environment and experience in these matters has been diverse, with some few similarities.

They come together with spontaneous circumstantial ease on a common grief or problem, hard work or loss, or increase of responsibility. This matter hits a responsive cord in both.

Mercury will develop this as long as it fits in with his primary life patterns. His maximum expression of responsibility is in an area apart from Saturn.

In the examples, the psychologically mature people are able to cope with their difficulties and restraints with system and perseverance.

In one case, a woman had Mercury N parallel Saturn of the man she married. They had a commensurate sense of equal responsibility that far exceeded selfish desires. During the marriage, Saturn took on a political post of serious importance and intense labor. Mercury kept pace with him in setting a national standard of culture and dignity. Mercury's primary expression of work and responsibility was in her own area, but cooperative to Saturn.

When Saturn was assassinated Mercury carried the load of grief and tragedy with a self-discipline and capacity of management that was admirable.

In another long-term marriage a man with Mercury N parallel his wife's Saturn found in her a responsive cord to his own maturity and sense of honor. They had both known the unhappiness of problematical marriage, divorce and loneliness. Mercury's primary work and responsibility went into his job.

Because of a weak heart, he had to exercise discipline in his personal exertions.

This wife's planet Saturn held the aspect to the man's natal Mercury for life. They lived their lifetime together in a mutual security and safety. They worked hard for their gains, with the clear conscience and peace of mind that comes with the honorable fulfillment of duty, and their unselfish regard for each other.

In other examples, Mercury is unable to change the situation between himself and Saturn to a responsible level of maturity. In these cases Mercury withdraws in grief, disappointment or cold indifference.

The progressions show a period of time when the two have common problems or increase in responsibility.

In a long-term marriage, the wife's Mercury was N parallel her husband's Saturn for 9 years. Their responsibilities held natural differences, as Mercury was busy with the home and children, Saturn was engrossed in business labor. Saturn's business took much of his time, and Mercury developed more of her resources in areas apart from Saturn.

At the peak of the aspect, the couple were drawn close by a shared grief in the death of one of their children.

The following year they went into a commitment of debt in the purchase of a new home. In the 9 years, the two developed maturity but in separate areas of self-discipline. In that period the marriage had stability but not much warmth.

In another marriage, the couple had highly diverse work interests and maturity levels. They separated over bad debts, deceit and selfishness.

When Mercury's patterns were such that he regarded Saturn as a burden, he seemed to instinctively move away shortly before Saturn became involved in a loss or difficulty.

The Mercury N parallel Saturn is separative only to the extent that Mercury must fulfill his primary work and responsibility in an area apart from Saturn.

Mercury South Parallel Saturn

Mercury has a different life pattern from Saturn in work and responsibility, discipline and restriction. There are large areas of similarity that overlap.

They come together as a natural result of circumstances on a matter of a common grief or problem, hard work or loss, or increase in responsibility. This matter implies a tension that requires an effort to solve.

Mercury has a diversity with Saturn but their bond is such that he is impelled to direct his responsible effort toward Saturn in his goals of security and maturity.

In these examples, Mercury either presents Saturn with his problems, or gets involved with Saturn's problems.

This may be as inadvertant as the girl with Mercury S parallel her mother's Saturn. At various periods in her life, she had pneumonia, a broken arm, a broken foot, and a general lack of strength. Though her life patterns of work and responsibility were different from those of her mother, their maturity patterns overlapped. They were able to share each other's burdens with mutual support.

In a romance, a woman with Mercury S parallel the man's Saturn imposed on him her anxiety and security needs until he fled, reenforcing her pattern of rejection.

A man with Mercury S parallel his irresponsible friend's Saturn not only had a psychological lack of security in himself, but added to it by loaning money to the friend at a loss.

In one long-term example of a woman with Mercury S parallel Saturn to a younger friend, Mercury often imposed her selfish demands upon Saturn. When Saturn herself was in trouble, lonely and in debt, Mercury unhesitantly offered the help of her strength as well as financial aid.

In many of the progressions, Mercury is impelled to a selfish demand of Saturn. In various cases Mercury cost Saturn hard work, financial loss, grief and worry, and discredit.

The S parallel is unifying to the extent and in the area that Mercury must fulfill a stringent relationship with Saturn while maintaining his own primary life pattern of work and responsibility.

Only in the examples where Mercury is making a responsible effort to maturity, does the aspect have any solidarity. Even among people who love each other, their respect is strained by the Mercury S parallel Saturn.

Mercury Split Parallel Saturn

Mercury's maximum contact with Saturn is on a matter of mature responsibility, problem or loss, loathing or tragedy.

Each has a level of work and responsibility apart from the other. They have areas that are uninteresting or incomprehensible to the other.

They can have a unity in ratio to their equality of mature self-discipline, when they cooperate with candor and respect.

They tend to be forced apart in direct ratio to their hidden areas of selfishness, greed, underhanded deceit or immaturity. An outside force of tragedy or discredit may interfere from that area of their separate lives.

The exact effect of the split parallel must be determined in collaboration with the individual charts, as it involves the entire separate maturity patterns of the two people.

The Mercury split parallel Saturn is a serious relationship. In some examples Mercury had instant antipathy to Saturn as their patterns of responsibility did not include each other.

Two persons both had Mercury split parallel Saturn to a man who cheated them of money.

In various cases, one or the other caused a worry, grief, or rejection of the other.

In some examples, the two had a close relationship in family and in friendships where their maximum contact was at a time of loneliness, loss and difficulty. With mature responsibility they helped each other through this period.

About one half of the progressions are indicative of a serious problem. In various cases, Mercury was abandoned, in debt, ran away from home, had a mental breakdown, was overworked, or died. Saturn was afraid and alone, overburdened, jailed, abandoned, rejected, discredited, killed, in debt, or in poor health.

In a harsh example a woman had Mercury split parallel to Saturn of the man who raped her daughter. The impact of this maximum contact between them caused shock, grief, and loathing. The legal difficulties were costly in money, time, and emotional insecurity. The man was imprisoned.

In a more typical example, a young woman had Mercury split parallel Saturn of her mother. The girl was restricted at home and antipathetic to her mother. The mother was in a period of debt and unhappiness in her personal life. During this period of maximum strain between them, the girl left home. Later when the aspect moved out of orb they had a familial contact again that was warm, and more mature than formerly.

A father with Mercury split parallel Saturn to his son loathed the child. He was jailed after beating the two year old boy. (The father also had Mercury square the child's Mars.)

In the least discordant examples, both persons have some nature of stress.

Two women had the Mercury split parallel Saturn for 13 years, who were devoted and loyal friends. Both were hard working, responsible people. Mercury had a long period of emotional and financial insecurity, during which she leaned heavily on Saturn's strength of maturity. Saturn on several occasions had had falls in which she had some injury. Both women had periods of discredit and enmity from other people. Their relationship with each other remained firm in respect and understanding.

One man and woman began a long-term love affair during this progression. They were discredited because of their age difference and because of social inacceptability toward their cohabitation without marriage.

Another couple began a marriage under the handicap of poverty and debt. Both had jobs, but it was a period of struggle to adjust to the responsibility.

The Mercury split parallel Saturn is often a difficult aspect. The persons who were troubled the least, were those who met their obligations with a mature strength of perseverance and a steady maturity.

This Chart Comparison illustrates a casual advantage and rejection in a short term relationship. There is no Mercury to Mercury aspect, natally or by progression.

	Eve to Brad		Brad to Eve
Natal ☿ 15♌12 20 N.16	∠☉ 2° ⊻♇ 1°	Natal ☿ 23♎20 12 N.15	N.♇ ♀ △♄ 6° ☍ASC 5°
Progressed ☿ 25♍30 16 N.10	☌♃ ☍♄	Progressed ☿ 12♎15 ℞ 2 N.20	∠♀ □♂

Eve and Brad met at a swinging costume party.

(Mercury N parallel Venus, natal; semi-square Venus, progressed.) Brad joined the group around the piano and began to make small talk with Eve. After dancing together, they went out on the patio. Eve was warmly responsive to Brad's approach to love making and he asked her to go home with him.

She agreed, but later when he looked for her, she'd left with someone else. Brad was irritated and critical, but impelled by the attraction he felt to find out her full name and phone number.

(Mercury semi-sextile Pluto, natal.) Eve had felt the response to Brad's N parallel Venus, but wanted their cooperation to be on her terms.

(Mercury square Mars, progressed.) Brad found her aggressive sexuality provocative and he made the effort to meet Eve again.

(Mercury conjunct Jupiter, progressed.) Eve had received a small inheritance that she was spending on good times and adventures. She acknowledged and appreciated the fact that Brad was a financially solvent businessman. They had several evenings of expensive entertainment together.

(Mercury trine Saturn, natal.) Because Eve was playing a frivolous game, Brad had no awareness of her capabilities. His objective was primarily that of conquest.

(Mercury opposition Asc, natal.) It was not difficult for Brad to reach an intimate relationship with Eve, even though their basic personalities and attitudes did not blend.

(Mercury opposition Saturn, progressed.) There was neither construction or responsibility to the contact. Within a week of their meeting, Eve and Brad had reached the extent of whatever relationship was possible to them under these circumstances. Their similarity was in a self-satisfying use of each other.

(Mercury semi-square Sun, natal.) Eve had no depth of understanding of Brad's character or attitudes. When he did not call her again, she felt critical irritation for a short while before the whole affair lost importance to her.

(Mercury in no aspect to Mercury.) During their week together, Eve and Brad never did reach any understanding communication. They had little to talk about of common interest or experience.

CHAPTER XI

Mercury in Aspect to Uranus

URANUS RULES INDEPENDENCE AND ORIGINALITY. It is the unique self-expression of individualism.

Mercury makes contact with Uranus on these matters. The aspect implies changes, either in the relationship or in the life patterns of the two people.

There are sudden events, new people and experiences that are radical departures from former patterns. Unconventional attachments and viewpoints have a magnetic effect.

The Mercury-Uranus parallel is a compelling exchange of different worlds between two people whose life patterns have a unique quality.

In all the Mercury-Uranus aspects, the one who is affected the most must be delineated from a study of the two separate charts with their progressions, to determine which one has Uranus in aspect at a given time.

Mercury Conjunct Uranus

Mercury has a familiarity with the individualism of Uranus. He understands or identifies from his own personal experience with Uranus' independence or originality.

They come together to share, or experience at the same time a unique experience that is sudden and unexpected, or a changing condition that will represent a new chapter in their lives. Mercury may have an impartial acknowledgement of Uranus. In the examples of long-term relationship the two people have a changing experience together at given periods.

In the case of a father with Mercury conjunct his daughter's Uranus, they made many moves together that were radical changes, from one country to another into entirely new conditions and environments.

A woman with Mercury conjunct her husband's Uranus went through a three-year period of change with him that began when they separated. With an independent perversity, he bought a business that was competitive to hers. Within six months they reconciled and moved into a new home that they decorated in modern style, a radical departure from their former way of life. Two years later, Mercury, the woman, had a drastically unexpected event when her business burned to the ground. She sold the acreage to an industrial firm that paid her twice the amount of the business value, and reinvested with her husband.

The changes are sometimes more subtle, as the two people's paths may cross at a time when they both are going through a period of independence or some unconventional pursuit.

In various cases, Mercury or Uranus had a nervous breakdown, and an explosive separation in his life, began a career in a field new to him, began occult or astrological studies, was jailed for unconventional conduct, had unusual personal disaster, suddenly went into public appearance, gained sudden prestige or great change of fortune.

The progressions show a specific time of change. Both persons go into a new chapter in their lives, together or independently of each other. They may move, change their status in some manner, form or break attachments, take up unconventional views or have sudden events occur that are unexpected.

As the conjunction is neutral, it does not bring a change of relationship directly between the two, unless other aspects influence this.

Mercury Opposition Uranus

Mercury and Uranus have in common an unconventional interest or situation. Where they have an understanding or identification on this matter, the full expression of their individualism does not blend.

The relationship can remain constant when held on a level of originality, where both persons accept the independence of the other. There is at some time a separation of unconventional viewpoint or a sudden experience of change. If that change involves a struggle for supremacy of the two independent individualities it will bring about unexpected separations that shock the tenure of the relationship, until one or the other disassociates.

There are several long-term relationships among the examples that were set in unusual conditions. A Korean woman with Mercury opposition the Uranus of an American soldier married him and left her home and country to come to the United States. His career took them all over the world and often separated them for months at a time.

Another couple lived together with the unconventional view of complete integral freedom of individuality. They both had sexual relationships and vivid communications apart from each other as well as together, with understanding and identification in their unique relationship.

Other long-term relationships and family associations had a common basis of radical interests and respect of individualism. One or both persons was in a unique field or had unconventional interests.

Often they came together at a time when one or both made a drastic change in their life's experience.

In a few cases the independence and originality of the two made an explosive clash of unexpected eccentricity.

In one case of progressions, a man had Mercury opposition his daughter's Uranus at the time when he had a mental breakdown and went berserk. She had to have him committed to an institution. (She also had Mercury S parallel his Saturn.)

In other examples, the two people were dynamically attracted to each other during changing periods in their lives. They may influence this change in each other; one may open new doors to an exciting world for the other; or one may unexpectedly display unconventional conduct and they disassociate.

Mercury Square Uranus

Mercury is strongly drawn to the unique quality of Uranus. There is an unexpected obstacle between them, or an incompatibility of independence.

When Mercury builds an association on the exchange of new ideas, radical interests and unconventional sciences there is a dynamic exchange.

When Mercury is eccentric or independent the relationship is erratic.

The Mercury square Uranus indicates at some time an unexpected change of relationship or an unconventional contact.

In the long-term examples, there is at some time a drastic change that is sudden and unexpected.

It is not always unusual except to these people. A man with Mercury square his girl friend's Uranus conceived a child in her. They married, but their conflict of independence led to separation.

In other examples, one or the other had periodic changing patterns that affected their relationship. In these various changes, one or the other had accidents, were murdered, jailed, had eccentric patterns, had sudden prestige or wealth, or was in an unusual occupation or situation.

There are examples of homosexual and bisexual relationships. There are examples of eccentricity to the point of mental aberration.

The progressions show a specific time that may represent a crisis. This can be beneficial in spite of the great tension, or it can be drastic.

A man made an acquaintance to whom his Mercury made the square to Uranus. The unique character quality of Uranus influenced Mercury to change his life patterns into a study of an unconventional science. Uranus himself was in a new chapter of his life's experience.

A man had Mercury square the Uranus of his young child at the time he made a critical change in his work. The child felt this only to the extent that he felt his father's tension.

In all the examples the aspect is felt by both persons to some extent, if only in their independence, or magnetic attraction to each other.

In some examples, the two people had sudden and drastic separations because of a move, a death in the family, or loss of work. Occasionally one or the other would disassociate in eccentric independence.

A few of the Mercury square Uranus examples are of stable relationships. Even these are subject to periodic explosive separations or changing patterns.

Mercury Trine Uranus

Mercury has a sudden and spontaneous contact with Uranus. Circumstances and environment are such that Mercury can have a beneficial change of relationship with little effort.

When Mercury is unappreciative of the unique quality of Uranus, the two have a passive independence.

When Mercury develops a relationship built on new ideas and radical changes, he enhances the individualism. Both have the good fortune of opening new realms of experience and change.

This is an exciting aspect. Mercury may step into new worlds of ideas and exchange with ease. At the same time he can have the adventure of introducing new conditions into Uranus' life. Often the two people go together into a new chapter in their lives.

Their common interest may be astrology, the occult sciences, radical movements, or unusual fields of experience.

Sometimes they meet under unusual circumstances. A man with Mercury trine the Uranus of his future wife met her when he was hitch-hiking across country and she picked him up. Their ten years together were an adventure of changes. They moved from apartment to house trailer to mansion to farm.

Another couple met on a railroad train, and fell radiantly in love. During their marriage they were able to exchange their original individualism, as his work was in the entertainment field and she was a writer.

There is usually a magnetic attraction between the two based on their recognition of unique interests in common.

In a few examples, other aspects contributed to antipathy. There were a few short lived love affairs and friendships when Mercury's independence was such that he made no effort to acknowledge the quality of Uranus.

The progressions show a specific time of exchange in the affairs of the two, or in their relationship with each other. There were marriages and love affairs, and people who met with magnetic attraction due to the radical interests they had in common.

A youngster with Mercury trine her mother's Uranus went on a trip with her to a foreign country.

A father with Mercury trine his daughter's Uranus was reunited with her after a year's separation.

A woman had Mercury trine her child's Uranus when the girl was two years old. The girl was a TV performer, whose magnetic charming appeal was such that she was pictured on a national magazine cover.

There is some unique quality or condition that is highly beneficial to Mercury, in all cases of Mercury trine Uranus.

Mercury Sextile Uranus

Mercury has the ability to change his relationship with Uranus in a sudden and unexpected manner. There is a certain time when the two are drawn together by new ideas, radical interests, or an unconventional attachment.

If Mercury does not choose to develop this, the time passes and the chance is lost.

In some of the examples the sudden and unexpected change was drastic, as one or the other was murdered or in a fatal accident.

There were many sudden and dynamic love affairs that were either set in an unstable framework, or begun during a period of change. The aspect is conducive to romantic attachments but not to endurance.

There are long-term friendships based on new ideas, radical and unconventional interests. These were subject to some periods of change, primarily because these were unique people. Their changes were in originality of work, home and hobbies, or change of tenure in the relationship itself.

In the progressions, several times Mercury flatly refused to have anything to do with Uranus, as he felt that Uranus was some kind of freak. Even when Mercury himself was an unusual person, he found something objectionable to Uranus.

A woman had Mercury sextile her husband's Uranus when he had a fatal accident.

In other examples the aspect had as much excitement and zest as the trine, where the two people regarded life and each other as an adventure.

Mercury Semi-sextile Uranus

Mercury has a vivid spark of interest in the unusual quality of Uranus. They can develop an unconventional attachment if both persons put an effort into this.

A man with Mercury semi-sextile the Uranus of a hippy boy took an interest in his unconventional way of life. As other aspects contributed to their affinity, they had an intermittent contact of friendship.

In a long-term relationship, there is at some time an unusual minor circumstance or condition between the two. In some cases there was no more than the spark of magnetic appeal.

In the progressions, when other aspects added weight, this was a drastic trigger. A girl with Mercury semi-sextile a man's Uranus, was attacked by him and injured. In a homosexual cohabitation the two broke up at this time.

In another relationship a woman cheerfully took a young man into her home who was a friend of her hippy son.

Two young men had a brief association together on a protest march of the Viet Nam war.

In other examples of the Mercury semi-sextile Uranus, the two people shared some unconventional pursuit such as going to freak-out parties, or unconventional gatherings.

Mercury Semi-square Uranus

Mercury's contact with Uranus has intermittent periods of sudden and unexpected change. Mercury does not fully understand the independence or unusual interests of Uranus. This may cause him some worry or anxiety that easily becomes criticism or irritation.

Mercury can handle the slight unexpected changes in their relationship with firm patience, as conditions either right themselves or lose importance.

In some of the examples the contact itself was intermittent at unexpected times. Mercury never knew when he'd have a chance to see Uranus. When they would get together they'd do something unusual in entertainment or adventure.

In a long-term relationship Mercury was worried and critical of her Uranus husband's periods of independent disassociation.

Another Mercury wife had a great deal of anxiety over the mental illness of her Uranus husband.

In one case, a mother had Mercury semi-square the Uranus of her son. When he was five, he began to work in television. This put them both into an entirely new environment. Mercury had anxiety over the boy's capacity.

In all cases of Mercury semi-square Uranus, the matter either stabilized in time or lost importance.

The progressions again held examples of intermittent radical interests between the two people.

Mercury Sesqui-square Uranus

Mercury has a sudden and unexpected situation, or recurrent intervals of disruption with Uranus. This implies a change of relationship that is highly agitating. Present conditions are radically broken up.

Mercury can make the decision of whether to begin, or continue, an unconventional relationship, or whether to let the matter drop.

In a few cases Mercury did let the matter drop. A man with Mercury sesqui-square the Uranus of an attractive woman was magnetically attracted to her unique quality, but the fact that she was Negro and he was white caused him agitation.

In other cases the two formed a dynamic friendship based on new ideas and radical views. Their recurrent periods of disruption were in the excitement of their periodic contact.

Several couples began love affairs while married to other people, causing a great deal of agitation. In one case Mercury broke off the affair; in another, Mercury carried it through to a disruptive marriage with her lover Uranus.

In a long-term family relationship, the disruption was in their nomadic existence of 18 moves in 10 years.

The progressions show a marked time of disruption. The sudden situation in various cases was a death in the family; Uranus was imprisoned for attack, leaving a Mercury friend; several love affairs and a marriage begun under drastic obstacles. Mercury is often magnetically impelled toward Uranus, even when he can forsee the possibility of agitation.

Mercury Inconjunct Uranus

Mercury is strongly attracted to the unique quality of Uranus. He takes the initiative in developing a relationship based on new ideas, radical and unconventional views.

In time Mercury's interest disperses into other areas. He can maintain an attachment for Uranus.

These people met with a dynamic attraction. In some cases it dispersed quickly, where there was a drastic exchange of independence or eccentricity.

In other cases there were long-term relationships of marriage and friendship, based on unusual qualities of originality and frequent changes.

A married couple moved continually, and world-wide. Several friendships were between a homosexual and a straight person.

The progressions indicated a climax of change that was separative. A woman with Mercury inconjunct the Uranus of her dear friend saw him for the last time. A year later he died a natural death.

A young man with Mercury inconjunct his mother's Uranus left home to go into the Army. He had the progressed aspect repeat itself 18 years later when he was killed in a plane crash.

A homozexual relationship broke up at this time.

In all the Mercury inconjunct Uranus examples both persons have some greater or lesser degree of change in their lives at the same time or together.

Mercury North Parallel Uranus

The changing patterns of Mercury's life differ from those of Uranus. Their previous environment and experience have been diverse in their unique self-expression of individualism.

They come together with spontaneous circumstantial ease. Mercury's independence and originality hit a responsive cord in Uranus.

Mercury will develop a relationship with Uranus as long as it fits in with his own patterns of change. His primary expression of individualism is in an area apart from Uranus.

When Mercury does develop a relationship with Uranus, it marks a new chapter of change in one or both lives. That change may be so dynamic as to alter their entire future from prior indications. We have that exemplified by a patient with Mercury N parallel Uranus of his plastic surgeon. (The doctor had Mercury opposition the patient's Mars.)

The change is often unequal. One person's eccentricities, or unique quality of character or experience may have a striking affect on the other. The change is not always together, but if not together, near the same time.

In all cases Mercury's primary expression of independent thought or action is in an area apart from Uranus.

A number of love affairs were sudden and dynamic. In each case there was something unique in the relationship. One couple were continually high on drugs. Another couple lived as man and wife who never were legally married. In one case of cohabitation, the man was heterosexual, the woman was bisexual.

In the case of a man who molested a child, they both had Mercury N parallel the Uranus of the other.

Some of the examples were of firm friendship based on new and radical ideas that contributed to both person's changing life patterns, where the contact was intermittent.

The progressions show a specific time of change for both persons. The new chapter may not be equal for both, but it does imply sudden events, new people and unusual experiences for both.

A woman had Mercury N parallel the Uranus of a homosexual man. Their relationship with each other was a pleasant friendship in which they occasionally had dinner together, and met in the same social group. At the time of the progression the woman broke up with her husband and went back to school. Within months of this, Uranus broke up with his boy friend and unexpectedly began going with a girl.

In other examples, the two people moved together into a new house or area; both changed their fields of work at the same time; or both defied convention in some radical manner.

The Mercury N parallel Uranus is separative to the extent and in that area that Mercury must fulfill his primary life pattern of unique individualism in an area apart from Uranus.

Mercury South Parallel Uranus

The changing patterns of Mercury's life differ from those of Uranus. There are large areas of similarity that overlap in their unique interest of individualism.

They come together as a natural result of circumstances that create a tension. Mercury has a diversity with Uranus, but their bond is such that he is impelled to a relationship that marks a chapter of change in his life.

In the long-term examples, the unique quality of the relationship may cover years of time. A child who was mentally and physically crippled had Mercury S parallel her mother's Uranus. She had periods of sudden and drastic changes in her deterioration of health.

In other examples Mercury was impelled toward the unique quality of character that Uranus had. In one such case Mercury had a tension of striving to fulfill the ethical concepts that Uranus embodied as spiritual teacher (Guru).

In family and friendships the tension, in some cases, developed to a certain point of crisis that marked a change in the relationship.

A man with Mercury S parallel his friend's Uranus, would turn to this friend whenever his unconventional life had another radical and unexpected change. The friend would respond with help and assistance. Mercury would quite involve Uranus with his unusual ideas until they both became nervous and frayed. At this point they would explode, and separate for another period of months or years.

The progressions show a specific time when Mercury is changing his life patterns. At the same time he is impelled to develop his relationship with Uranus. Sometimes the two events coincide, as when Mercury and Uranus marry.

In other examples Mercury may be going into a totally different chapter of experience while maintaining a positive active contact with Uranus.

The Mercury S parallel Uranus is unifying to the extent and in that area where Mercury must fulfill a dynamic attachment with Uranus, while maintaining his own independent individualism.

Mercury Split Parallel Uranus

Mercury's maximum contact with Uranus is on a matter of sudden and unexpected change that marks a new chapter in their lives.

Each has a level of individualism or unconventional interest apart from each other. These areas are uninteresting or incomprehensible to the other.

They can have a cohesive unity in candor and cooperation that gives respectful recognition to each other's independence and originality. They tend to be forced apart in direct ratio to covert or coercive eccentricities or erratic patterns.

A sudden or unexpected event may interfere from that outside area of their separate lives.

Surprisingly few examples of the Mercury split parallel Uranus have drastic changes. The majority of examples are of people who have some unusual quality or character in their personality, work or interests. In long-term relationships, they meet the changing events of their life patterns as a new adventure that is stimulating.

Mercury and Uranus, in one case, were of different nationalities, and married in a country foreign to both of them.

In another example, Mercury and Uranus had a conventional background and marriage and had triplets, three baby girls.

In other cases their changing patterns separated the two people then brought them back together again.

At times they had sudden and unexpected events and radical interests apart from each other. When they respected these separate areas of each other without imposition, they had a harmonious unity that enhanced their individualism.

In one of the few cases of drastic change, Mercury and Uranus had a son who committed suicide at the age of 32.

In another, Mercury loaned a large sum of money to Uranus that was lost in a business venture (refer to Mercury split parallel Saturn.)

In the progressions there was a more marked maximum of contact on a unique level. A woman Mercury met a Japanese artist, Uranus, who introduced to her a new world of ideas.

A woman Mercury left her Uranus mother. A year later she married; the following year while the progression was still in effect, the mother remarried.

In most of the examples, the changing patterns of the two people are not directly together but they do influence each other.

In the Mercury split parallel Uranus, the two people offer each other unique unity in their diversity.

The Chart Comparison on page 150 illustrates a working relationship between two men under most unusual conditions. The chart is of two astronauts who manned a space craft.

Throughout the chart comparisons the true names have not been used. In some cases this was to protect personal information. Generally this has been to present the quality of interaction irrespective of identity.

The astronauts in this Chart Comparison are world-known figures. To maintain policy we will use initials.

The personal relationship of the two men is not the matter of interest in this chart comparison.

The two Uranus aspects indicate the unique quality of the individuals as well as their outstanding contribution.

The one Mars aspect indicates the disruptive conditions of pressure in their work. Combined with the three Saturn aspects we can judge the tension of responsibility. A critical danger would not be expected from a careful judgment of the aspects.

The total of four Sun aspects and three Mercury aspects show the exchange in communication and relationship based on a similar goal and interest.

Chart of Two Astronauts Who Manned a Space Craft

	W.S. to T.S.		T.S. to W.S.
Natal ☿ 29♒06 13 S.39	⚻ ☿ 3° ⚼ ♀ 4° ⚼ ♂ 2° ⚼ ♃ 2° ⚹ ♄ 6° ∠ ♅ 1°	Natal ☿ 2♎36 ℞ 4 S.14	⚻ ☿ 3° △ ♀ 5° s.P ☉ ∠ ♃ 1° s.P ♄ ∠ ♆ 1°
Progressed 1965 ☿ 17♉51 18 N.42	⚼ ☿ ⚻ ☿p	Progressed 1965 ☿ 18♎01 5 S.29	⚻ ☿p ⚼ ☉p ☌ ♄ s.P ♅
Natal ☿	△ ☉p	Natal ☿	⚼ ☿p ⚻ ☉p

CHAPTER XII

Mercury in Aspect to Neptune

NEPTUNE RULES THE IMAGINATIVE VISION of a perfect community of voluntary consideration. It is the sensitive dream of the aesthetic ideal, or the subtle secret fantasy toward ease, ecstacy and the erotic sensations.

Mercury makes contact with Neptune on these matters. There is something in the nature of the exchange that is idealistic, and something that is disappointing. The Mercury Neptune contacts are not entirely realistic in a practical sense, and often are exaggerated or hold a note of mystery, scandal, or schemes.

Neptune includes the mystical experience of visions, psychic ability and extra sensory perception. It rules the dramatic ability or proclivity. It rules flight, whether in an aircraft or in sensation.

The Mercury-Neptune parallel is a subtle exchange that includes secret areas in the larger context of the two separate life patterns of dreams and visions.

In all the Mercury-Neptune aspects, the full nature of effect must be delineated from a study of the two separate charts with progressions, to determine which person has Neptune in aspect at a given time.

Mercury Conjunct Neptune

Mercury has a familiarity with the ideals or fantasies of Neptune. He understands or identifies from his own personal experience with Neptune's sensitive dream of the aesthetic ideal, or with his secret fantasy toward ease, ecstacy and the erotic sensations.

They come together to share, or experience at the same time a quality or condition that is idealic, or that holds a note of mystery or scandal.

As the conjunction is neutral, Mercury can acknowledge Neptune impartially.

Often there is a full Neptunian expression in the contact. In a long-term marriage, a man had Mercury conjunct his wife's Neptune. They were well suited for each other, as they fit each other's ideals of pleasurable exchange, as well as having the innate ability to work well together.

Mercury was a lusty, virile man, who understood well his Scorpio wife's erotic needs. They had a communal voluntary cooperation in the construction of their business. When this business sold, the profit was so great that the two were able to take an extended vacation to Hawaii that was a long cherished dream.

There was one period during their marriage when they had a scandal. They were involved in a mate swapping club that was dramatically exposed to the public.

A woman who was a neighbor to this couple had Mercury conjunct the man's Neptune. She expressed her own imagination in the community little theatre group. She was not involved in the sex club but knew of its existence. She was able to accept the man impartially without any urge to personal involvement.

When Mercury and Neptune maintain an integral aesthetic ideal there is little chance of scandal.

A married woman had Mercury conjunct Neptune of a male friend who fell in love with her. She kept the relationship on an open friendship level in which they exchanged their deepest dreams, hopes and visions in a communication that approached the mystical. (They also had Mercury split parallel Mercury.) He did write her poetic love letters that she kept secret.

In some cases Mercury's impartiality excluded Neptune, and there was no more than secret fantasies between them.

The progressions are subtle. Often the exchange remains a secret to the public, where no more than a few people ever know the extent of what happened. There were secret love affairs that were ecstatic and erotic; and one case of secret abortion of a pregnancy.

Two couples belonged to a communal group that shared sex and dope. In other examples the two exchanged the mystical experience of aesthetic love and profound idealism. There was one relationship of this nature between a heterosexual woman and a homosexual man.

In a number of the examples one or both smoked marijuana or took hallucinogenic drugs.

A point will be clarified here. The example charts include approximately 60 persons who have had drug experiences. Neptune does rule drugs, and in the examples where drugs are a distinguished part of the exchange between these two people, it will be mentioned. The Mercury-Neptune aspect between people born in the early 20th Century is not apt to pertain to drug use as it was less common then. Any study reflects the time in which it is taken. An effort has been made to base the "Mercury Method" on as large a variety of human relationships as the records allow, of people from all walks of life.

The people who take drugs have Mercury to Jupiter aspects as this is often a new expansion of experience or indulgence. They have Mercury to Uranus aspects, as it is often a radical departure from convention. They have Mercury to Mars aspects when the drug use leads to sexual exchange, or rarely, violence.

The Neptune aspects indicate the aesthetic or ecstatic fantasies, the release of tension or effort, or the mystical experience that comes with drug use. These same reactions are experienced by many people who never take drugs.

In an example of a heroin addict, his wife had Mercury sextile his Neptune as the sole indication of their shared use of various drugs.

Marijuana and all the hallucinogenic drugs will be categorically referred to as drugs unless distinction is required in a specific case.

Mercury Opposition Neptune

Mercury has in common with Neptune a utopian vision, either of relationship or situation. They reach an understanding and identification on a certain level that is idealic for both.

The full expression of the matter does not blend in a practical sense. When an adaptation to reality is required there is at some time a separation of ideals or a disappointment.

When Mercury and Neptune have too wide a diversity, as when one is on an erotic fantasy level and the other on an aesthetic ideal level, they clash and separate, sometimes with revulsion.

When Mercury has the identification and understanding of Neptune's level as being similar, they develop a common situation.

One marriage, in which the woman's Mercury was opposition the man's Neptune, was based on their common ecstacy in deviate sexual exchange. On certain occasions during their years together they had bitter clashes of suspicion and deceit.

Another romance was based on idealic trust and an aesthetic purity of love. These two people could not fully blend into a practical relationship of marriage because of circumstantial prohibition. Nor did they feel that they could consummate a sexual fulfillment because of their morality ideals. They eventually separated, each with the memory of the perfect first love.

In other examples Mercury had an exaggerated image of Neptune and entertained secret fantasies that were inevitably disappointing when he attempted to bring them into reality. There was a great deal of exchange in these relationships that was ideal for both, when kept on their level of agreement.

The progressions had one adventurous example of two men with Mercury opposition Neptune, who formed a partnership to go on a treasure hunt. The hunt itself was idealic with drama and mystery, as well as the camaraderie of communal effort. The practical reward of actually finding the treasure did not materialize.

There is a contact between two people, both married to someone else, who were involved in a sex club. As their relationship was based on half truths, concealed motives and erotic lust, in a situation that was potentially scandalous, the two separated in exaggerated emotional confusion.

In all the Mercury opposition Neptune examples an idealic relationship can be maintained on a level of aesthetic integrity. Where there is deviation, deceit, schemes, or unrealistic exaggeration, the two are bitterly forced apart.

Mercury Square Neptune

Mercury is strongly drawn to an exaggerated image of Neptune. There is an obstacle in reality, or an incompatibility in their ideals.

When Mercury has the integral sensitivity of voluntary consideration, he can develop an idealic exchange with Neptune.

When Mercury has subtle secret fantasies, or bases the relationship on an unrealistic view, he creates a confusion or disappointment through which the contact is lost.

Mercury's image of Neptune may be exaggerated in either over or under estimation. He may reenforce either by his own standards.

In several cases, Mercury idealized friend Neptune. He related to Neptune with such trust and voluntary cooperation that they did indeed have a relationship that was aesthetically ideal.

A young man had Mercury square Neptune of the girl he married. They had a radiant relationship, socially, emotionally, and sexually. Their obstacle was in the harsh reality of jobs, money and the necessities of worldly life.

Instead of prosaic work, the young man pursued an acting career. During the second year of marriage, they ran out of clothes and food and were living in a cold-water flat. Their utopia dissolved in a heart-breaking separation.

There is often a measure of disappointment between the two, generally when Mercury has an unrealistic evaluation.

A woman with Mercury square Neptune of a man not only idealized him but dreamed of a perfect match with him. This was not his intent at all, and they separated.

In another case, they married. Mercury maintained a fantasy picture of Neptune that was unrealistic beyond his capacities.

One woman with Mercury square her neighbor's Neptune fantasized in imagination Neptune's affairs beyond realistic truth. She created a minor scandal by saying that Neptune had loose sex affairs and drugs in the house.

In the progressions there were examples of contact that faded into subtle fantasy.

In one case a wise man, Mercury, with the sensitivity of psychic vision made the effort to develop Neptune's ideals in a manner that insured their devoted bond.

In a scandalous case, two homosexual men had Mercury square Neptune at the time when Mercury was rushed to the hospital from an overdose of drugs.

In many of the examples, one or both persons were in Neptunian occupations or pursuits. They took drugs, were in theatrical fields, art or music, or the priesthood.

Mercury Trine Neptune

Mercury has a spontaneous contact with Neptune in which he easily offers some idealic exchange. Circumstances and environment are such that with little or no effort Mercury can stimulate the imagination of Neptune.

When Mercury stimulates the subtle secret fantasies of Neptune he may lose the benefit entirely.

When Mercury presents his dreams and ideals on an aesthetic level of integrity that is realistic, he has the benefit of a voluntary consideration with Neptune.

The Mercury trine Neptune has examples of association that were as close to perfection as humanly possible. In these cases Mercury set the standard and made the effort to share with Neptune his dreams and ideals of aesthetic integrity. He was at the same time practical and realistic. There were persons who exchanged a relationship of mutually devoted consideration and trust. They had a joy and delight in wildly imaginative ideas; they shared music and art and drama.

The planet Neptune is so subtle that in some cases Mercury had no more than a vague idealized picture of Neptune that he made no effort to develop.

One couple met at a vacation resort. Mercury was a man who saw his ideal of perfection in a mirror, and whose imagination extended no further than sex fantasy. After one date, Neptune found him revolting.

The progressions have a subtle quality. Often the two are able to exchange an idealic consideration of each other's sensitive dreams. Mercury easily understands Neptune's vision and may encourage his music, art, psychic or aesthetic goals.

In an unfortunate incident, a young girl had Mercury trine the Neptune of a male neighbor. With her budding sensuality of adolescence she inadvertently stimulated the secret fantasies that he satisfied in voyeurism. There were several weeks of mystery when footsteps were found outside of her bedroom window, and a brief scandal when he was caught.

In the most outstanding example the dreams became a practical reality of two men with Mercury trine Neptune. They based a corporate partnership on the vision of a better product. With their voluntary cooperation to an ideal goal, they were able to get financial backing of one million dollars.

The exaggeration of the planet Neptune is such that many times it only gives a fraction of what it promises. On other occasions, Neptune is the pot of gold at the foot of the rainbow, and gives beyond the wildest dreams.

Mercury Sextile Neptune

Mercury has the ability to see the ideal quality of Neptune. There is a certain time when the two are drawn together easily and the opportunity is presented for an exchange, in a matter that is mystical, ecstatic, or aesthetically idealic, mysterious, scandalous, or strange.

If Mercury does not accept this opportunity at the time that it is offered, the chance passes like a dream.

The exchange can be on any Neptunian level. With the wide variability of the sextile, these examples hold every possible expression of the planet's nature.

In various cases, the interaction was mystical and psychic; a business scheme that became confused and exaggerated; a mysterious death of Mercury; a mental breakdown of delusion by Mercury; a secret love affair; an erotic love affair; drug use; and persons with experience in music, art, and drama.

The progressions have the same type of affair, at a specific time. Mercury has the opportunity presented for some quality or condition that he finds ideal.

Often the subtlety of the aspect is such that it remains a secret to the public. There may be a vague note of threatening scandal; secret dreams and illusions, mysterious and muddled affairs. Mercury may refuse a contact with Neptune for these reasons.

Their exchange is not always on an equal level. In one example a young man had Mercury sextile Neptune of an older woman who was a dear friend. At the time of the progression he was producing his first movie, a fantasy set to music. She was pursuing her private dream in psychic research.

In another case Neptune was murdered. Mercury took the opportunity to eulogize his memory in idealistic terms.

The Mercury sextile Neptune shows the opportunity, no more. The outcome and effort is a matter of the choice and effort of the two people.

Mercury Semi-sextile Neptune

Mercury has a vivid spark of interest in the subtle quality of Neptune. With combined effort, they can develop a relationship that is aesthetic, erotic, or idealic.

When other aspects are prohibitive, this is no more than a mild stimulant. In other examples, the Mercury semi-sextile Neptune was the spark to set a tone of exquisite relationship.

In various cases, the two had a love based on voluntary consideration. a psychic exchange; an erotic affair; one or both had experience in drugs, theatre, aircraft, or homosexuality.

Mercury Semi-square Neptune

In Mercury's contact with Neptune, he has intermittent periods of worry or an anxiety of complete confusion when he finds Neptune's position or attitude incomprehensible. Mercury may be critical or irritated by the vague change in their contact.

He can be firm and patient, as conditions right themselves or fade in importance.

In the long-term relationships, there are occasions when Mercury has no idea of where Neptune is, physically or mentally. A woman with Mercury semi-square her husband's Neptune was periodically concerned when he would disappear for days at a time. She could only fantasize his erotic or aesthetic levels as he would not discuss the matter.

Another woman with Mercury semi-square her husband's Neptune had a vague suspicion that he was involved in an illegal scheme. When she found that this was true she thought it incomprehensible. It lost importance when the affair fizzled.

A man with Mercury semi-square his wife's Neptune often had anxiety over certain psychotic patterns of hers.

In other varying cases Mercury worried about Neptune's drug experiences, business schemes or instability. In some examples, Mercury expressed a vague confusion in that he had no idea of what Neptune was doing or thinking.

Mercury Sesqui-square Neptune

Mercury has a sudden situation with Neptune of mystery or scandal, or inexplicably muddled conditions at recurrent intervals, that breaks up present conditions. Mercury may make the decision to begin or readjust a contact on more idealic terms.

In a typical example, two men who were friends, had Mercury sesqui-square Neptune. When they met, Neptune was an out-of-work designer. Mercury, in voluntary consideration, offered Neptune a place to stay.

Mercury admired Neptune's artistic quality of work and they shared a pleasure in classical music. They continued to live together, with occasional periods of disruption in conflicting patterns. Mercury was always late. Neptune had several contracts of design for which he received a reputation of brilliance, but his financial affairs were always muddled. Mercury had a love affair with a married woman that became public, causing discredit. In a clash of moral ideals he and Neptune separated.

In another case, a woman had Mercury sesqui-square her husband's Neptune. She made the decision to maintain the ideal quality of their marriage after he had a health problem that rendered him impotent.

In the progressions, a man had Mercury sesqui-square Neptune of his wife and daughter both at the same time.

He was found dead in the street with two skull fractures. The cause of death remained a complete mystery.

In other examples there is some vague impossible confusion between the two or a mystery as to what was the nature of their relationship.

Mercury Inconjunct Neptune

Mercury is strongly attracted to an idealized image of Neptune. He takes interest and initiative in developing a voluntary consideration. Mercury's efforts disperse into other areas, but he can maintain a certain level of ideal exchange with Neptune.

These examples include two of the best marriages on record, and several of the finest friendships. The two shared their sensitive dreams of aesthetic ideals, and an exchange of consideration.

In a few cases the two also shared a secret of one or both. That secret was their love affair, a psychic experience, a drug experience, or homosexuality.

The progressions indicate a time when the two are very close. In most of the cases, one or both reach a climax of some ideal, or Neptunian condition.

This may be the ideal of friendship, of psychic experience, of theatrical prestige, or the realization of a dream. The condition can be maintained, but not at its climactic level.

In the other examples the climax was secret, scandalous, or mysterious.

A woman with Mercury inconjunct Neptune of a girl-friend responded to her with consideration and devotion when

Neptune was raped. (The planet Neptune itself does not lead to rape, as it is gentle in nature. The two women also had Mercury split parallel Mars, maximum contact on a matter of violence; and Mercury south parallel Venus, impelling response to an emotional situation. The Mercury inconjunct Neptune is applicable to the fact that the affair was kept secret.)

In the muddled examples, the matter dispersed with some vague confusion.

Mercury North Parallel Neptune

Mercury has a different pattern from Neptune in his utopian urges. Their previous environment and experience have been diverse, in their expression of aesthetic ideals, mystical experiences, or illusory fantasies, with some few similarities.

They come together with spontaneous circumstantial ease on a matter that is idealic, mysterious, or scandalous. This matter hits a responsive chord in both persons.

Mercury will develop this as long as it fits in with his primary life pattern. The maximum expression of his dreams is in an area apart from Neptune.

There is often a strange, subtle quality to this contact. Because of their diversities, the two may not fully know where the other person is, in their separate, secret worlds.

Several of the examples of Mercury N parallel Neptune were highly mystical, in which there was an exchange of the extra sensory perception of telepathy. In the example of teacher (Guru) to disciple, there was the responsive chord of the highest ideals to integrity. In each case Mercury was expressing the practical application of his ideals in an area apart from Neptune.

In one secret love affair, Mercury disappeared. One day she was simply gone. Neptune never knew where, or why.

In another example of Mercury N parallel Neptune, two women friends worked together in a motion picture studio. They had an ideal communal exchange in their working environment, but Mercury did not know that Neptune was secretly married; Neptune never knew that Mercury had a prison record.

In another scandal, a woman had Mercury N parallel Neptune to the man who molested her daughter. Mercury said, "The same thing happened to me when I was a child!"

The Mercury N parallel Neptune examples also include two of the finest marriages possible, in the exchange of voluntary consideration. One of these was contracted during the progressed aspect.

In other examples of progressions, the two came together with a bond that was Neptunian in some way: in a sex club or bi-sexual relationship; in a hippy community; in drug use; in illicit secrecy; at a time of strange mental quirks; at a time of scandal; at a time of mysterious death; on a level of aesthetic ideals; at the time of a treasure hunt; in music, art, or drama.

In several cases, Mercury or Neptune was an entertainer, artist, musician or playwright. Whatever their bond, Mercury always had something else going on apart from Neptune, if only in his imagination.

In some cases their diversities led to a gradual separation. There are two examples of absolute flat antipathy. In both of these cases the basis of contact would have been erotic, and this was unacceptable to the other person.

The Mercury N parallel Neptune is separative only to the extent and in that area that Mercury must fulfill his primary life pattern of ideals, or illusions, apart from Neptune.

Mercury South Parallel Neptune

Mercury has a different life pattern from Neptune in his utopian ideals. There are large areas of similarity that overlap.

They come together as a natural result of circumstances on a matter of aesthetic ideals, mystical experiences, or illusory fantasies. This matter implies a tension of fascination.

Mercury has a diversity with Neptune, but their bond is such that he is impelled to an expression with Neptune that is idealic, mysterious, or scandalous.

The Mercury S parallel Neptune is unifying to the extent, and in that area, that Mercury must fulfill a meaningful relationship with Neptune while maintaining his primary life pattern of dreams, or illusions.

In 1944 Neptune moved into South declination, where it will stay until the year 2024.

We have very few example charts with Mercury S parallel Neptune. This age level includes many of the young people who are involved in the hippy communal living ventures, or the drug scene, both of which are ruled by Neptune.

We may hope that these people with Mercury S parallel Neptune will be impelled together by the tension of their ideals toward a better world that is a finer community based on consideration of each other.

Mercury Split Parallel Neptune

Mercury's maximum contact with Neptune is on a matter of aesthetic ideals, erotic or ecstatic experiences, or illusory fantasies.

Each has a level of idealism or subtle experience apart from the other. They have secret areas that are uninteresting or incomprehensible to the other.

They can have a cohesive unity by maintaining the utmost candor and consideration. They tend to be forced apart in direct ratio to covert, coercive or hidden schemes, or exaggerated romanticism. An outside force from that area of their separate lives can impose scandal, mystery, or bitter disappointment.

A few of these examples have a mute external effect; only on close examination can it be determined that there is a level of fantasy that the two people have, either toward each other or in their separate areas.

In some cases, the two had an integral ideal of practical consideration where there were no secrets. These were idealic relationships.

In many of the examples the maximum contact these two people had was in the bitter impact of scandal or mystery that involved one or both, or another situation. In a case of two women, they had little personal contact. Their maximum relationship was in the public scandal that occurred when a man divorced Neptune to marry Mercury.

One couple lived together for 15 years, where the woman had Mercury split parallel Neptune to the man. Both had strange and mysterious levels of experiences in their past, and secret hopes and visions apart from each other. They were involved in a number of business schemes through the years, that on occasion were quite profitable. The maximum of their relationship was reached when they went into a theatrical venture that was unrealistic and impractical. Neptune began an affair with an actress and Mercury left him.

In the progressions, the maximum level of contact occurred in secret, erotic love affairs of exaggerated intensity; in mysterious accidental deaths, from two of which Mercury received a large inheritance; and during periods of exaggerated confusion.

In three cases, the two people made fortunes of money, twice in business together, in the third instance separately but at a time of close contact.

Other examples are of people who had a mystical or aesthetic relationship, or who reached no more than idealized fantasy. The unity was impossible where there was no equality.

In the Mercury split parallel Neptune examples, where there is a maximum contact together, there are still large areas of separate experience that are difficult for the two to share.

The Chart Comparison on page 165 illustrates a psychic exchange. It is of a mystic to two of her clients.

Elsa is a Christian mystic whose quality of psychic perception has been examined by University tests and reported internationally.

The chart comparison is of Elsa to two women who visited her four times in the course of five years for readings. Both women took notes that they referred to over the year's time. Chris not only was given a longer and more detailed interview, but the extra sensory perception of Elsa into Chris's life was more accurate. Her predictions of the future were 80% correct.

Chart Comparison Illustrating a Psychic Exchange

Natal	Elsa to Chris	Natal	Chris to Elsa
☿ 11 ♎ 31 6 S.02	△ ☿ 8° ⚹ ♄ 5° ☍ ♅ 3° ∠ ♆ 1° △ Asc 2°	☿ 20 ♊ 22 20 N.20	△ ☿ 8° ⚺ ☽ 1° ⚻ ♅ 2° ⚺ ♆ 1°
Natal	**Elsa to Dawn**	**Natal**	**Dawn to Elsa**
☿ 11 ♎ 31 6 S.02	⚹ ☿ 2° △ ♀ 1° ☍ ♄ 6° □ ♃ 10° ⚹ ☽ 1°	☿ 8 ♌ 46 20 N.00	⚹ ☿ 2° ∠ ♃ 1° ⚹ MC 1°

Dawn did not feel that it was a satisfactory experience, nor did she find it meaningful or accurate.

Elsa and Chris both have Mercury-Neptune aspects to each other. Elsa and Dawn do not.

The conclusion cannot be assumed on short evidence that the Mercury-Neptune aspect is necessary for a psychic exchange. The possibility can be considered for further research.

CHAPTER XIII

Mercury in Aspect to Pluto

PLUTO RULES THE HEIGHTS and depths of experience. It is the urge to cooperate for the general good, or to manipulate in coersion for self again.

Mercury makes contact with Pluto on these matters. It is an exchange of the deepest development of possibilities, or of the dark and covert will to rule or ruin. It usually implies hidden levels, subtle force, and some measure of inversion (turning matters inside out in a reversal.)

The Mercury-Pluto aspect has to do with the most integral spiritual aspirations, or with insidious, drastic force.

The planet Pluto rules any group activity that has a common goal.

The Mercury-Pluto parallel shows the compelling areas of cooperation or coercion of the two people in the larger context of their separate life patterns of aspiration.

In all the Mercury-Pluto aspects, the one who is coercive or cooperative, or the one who is in a period of transmutation or metamorphosis, must be delineated from a study of the two separate charts with progressions, to determine which person has Pluto in aspect at a given time.

Mercury Conjunct Pluto

Mercury has a familiarity with the heights and depths of Pluto. He understands or identifies from his own personal experience with Pluto's ethos, his aspirations, or with his insidious dark areas.

They come together to share, or experience at the same time a matter of cooperation for their mutual benefit (or for a greater general good) or a coercive experience.

As the conjunction is neutral, Mercury can acknowledge Pluto impartially.

Many of these examples are of people who make contact in a group situation. These were ethical groups, musical groups, or work teams, etc.

In every case one, or both, have some hidden heights or depths, from each other or from the world.

A young man had Mercury conjunct the Pluto of two women who made a profound influence on his life. Mercury was a resourceful, strong willed man whose innate inclination was to examine life from the inside out in a depth analysis.

One of the women was his mother. She was a dynamic Scorpio, whose approach to him from his childhood on was, "Do it, or else." Their relationship was primarily a clash of wills. They did love each other. With maturity, Mercury was able to acknowledge his mother with an understanding that led to cooperation.

Mercury had an area of spiritual aspiration, to dedicate his work to the evolution of man, that he did not feel able to share with his mother.

In his early years at college, he met the other woman to whom he had the Mercury conjunct Pluto. She was the doctor who headed the research lab group. It was she who fired his full passion of devotion to the general good. In cooperation with her, Mercury developed his own hidden capacities, and found his firm direction. He became an outstanding neurologist.

In another example, two people with Mercury conjunct Pluto had little personal contact. They cooperated in a mutual trust and benefit, in that Mercury rented Pluto's house.

Two love affairs were based on hidden motives and subtle force by Pluto. In one case there was an insidious threat.

In the progressions there were varying degrees of cooperation for a broad area of general good and mutual benefit.

In one example a man had Mercury conjunct his wife's Pluto after 15 years of marriage. Both were in a period of inverting their former patterns, he in his work, she in her expressions of self will. Mercury suspected Pluto of infidelity. He had her followed by detectives and he threatened her with an ultimatum. They stayed together but the relationship itself disintegrated.

In several cases, Mercury attempted a work and social cooperation with Pluto, but met a coercive experience of covetry.

In all the Mercury-Pluto aspects, the contact implies a time when both persons are in a period of intensity, or inversion of pattern, temporarily or permanently.

Mercury Opposition Pluto

Mercury has in common with Pluto certain heights and depths of experience. Where they reach an understanding or identification of this, the full expression of their spiritual aspirations, or their dark areas, does not blend.

They may attain, or maintain a contact on a certain level. When a larger adaptation of cooperation is required, an incidious coercion becomes irreconcilable. They must withdraw until they can again cooperate on a basis of mutual benefit.

A typical example is in that of two men friends with Mercury opposition Pluto. They often worked in group activities for a common goal, in the army, as musicians and in their church group. Mercury was able to identify with Pluto in both his highest aspirations and his coercive self will. Periodically they would develop an impasse of hidden conflicting force, or the urge to rule or ruin that both possessed. They would separate for months to years at a time, and come together again at the time of a subtle inversion of change in their life patterns.

A marriage with Mercury opposition Pluto maintained a cooperative level for many years until the deepest inner needs of the two began to conflict. There were hidden levels that the two were unable to exchange. Both persons had heights and depths that were incomprehensible to the other. They parted into different group environments, but still cooperated in their business settlement.

Among the examples there is the cooperation between a dentist and patient.

There is a homosexual relationship between two men. Their level of heights and depths, coercion or cooperation are not on record. They did separate after two years of cohabitation.

The progressions hold examples of drastic coercive force that separated the two, as well as examples of periods of intense cooperation.

A priest with Mercury opposition the Pluto of a young woman worked with her in church groups and personally supported her through a period of reversing her aspirations to a courageous and noble decision.

A woman had Mercury opposition the Pluto of her husband when he was involved in a gang whose hidden activities never were fully revealed. There was sufficient exposure to send him to prison.

In another example Mercury's father Pluto was murdered. The insidious criminal force of this event placed Mercury in the position of necessary cooperation with the police to investigate his father's hidden activities.

A man with Mercury opposition the Pluto of a young woman forced a drug on her, prior to sexual assault. (Refer to Mercury S parallel Mars.)

In some cases of Mercury opposition Pluto the two have a period of temporary separation of viewpoint or experience in a basically harmonious relationship because they are in a time of deepening or enlarging aspirations and experience.

Mercury Square Pluto

Mercury is strongly drawn to coercion or cooperation with Pluto. There is an obstacle of incompatibility in their ethical heights or covert depths.

When Mercury has the highest aspirations he can build a constructive relationship for the general good.

When Mercury develops a contact based on coercion, the subsequent inversion will leave nothing of value.

There are a few examples in which the two overcame obstacles of distance or diversity to cooperate for good in spiritual aspirations.

The mundane examples are less fortunate. There are hidden depths of coercion and psychological warp in one or both that meet in a crisis.

A long-term marriage was established on this basis. The man had Mercury square the woman's Pluto. Their romance began with Mercury's proposition, "We will make love, or else separate." Pluto later countered with, "We will marry, or else I will cause a scandal."

The hidden areas of decadence and destructive force in this relationship finally did lead to Mercury's psychological breakdown of competence.

The most harmonious relationship of Mercury Square Pluto has some nature of coercion in it. The premise is, "We will cooperate for this height, or depth, or we will separate."

A psychologist had Mercury square the Pluto of a patient.

A student had Mercury square the Pluto of his teacher in spiritual ethics.

A patient had Mercury square the Pluto of his doctor.

In the progressions there are similar examples and some that were less harsh.

A man with Mercury square the Pluto of his wife joined the church choir with her.

Another couple attempted to work in cooperation on a difficult project that they were unable to complete.

Several marriage and love affairs were broken up in a conflict of subtle self gain.

Mercury Trine Pluto

Mercury spontaneously falls into a cooperative contact with Pluto. Circumstances and environment are such that Mercury need make little effort to elicit a depth of exchange with Pluto for the common good of both.

When Mercury extends his integral high aspirations, there is a superior association.

When Mercury is indifferent, or covert, he may lose the benefit entirely.

The deepest and finest of human relationships for the good of both can be directed to the good of mankind. Mercury can tap the spiritual heights with Pluto.

There is one example of a relationship of which we can only speak in reverence, as Mercury was known as a saint among men. He and Pluto met in spontaneous devotion, and were blessed in their work together for humanity, in literature, art, and culture.

Other examples were more mundane, where the two had a common work, or cooperated in beneficial group activities.

Mercury Sextile Pluto

Mercury has the ability to tap the heights or depths of Pluto. At a certain time the two are drawn together easily and the opportunity is presented for their cooperation.

If Mercury accepts this chance, he can develop their unified force in mutual benefit for the general good.

If Mercury ignores this, the time passes and the chance is lost.

Several relationships were built upon this aspect in which the two people cooperated for the greatest good, in a depth of inner development and awareness. These contacts came from a spiritual group, a psychological encounter group, or from cohabitation.

In other cases, Mercury gave a mild cooperation to Pluto for a short period only.

In one situation, a man had Mercury sextile the Pluto of two persons in his car pool. This was cooperative, but Mercury's personal level was self will to the point of tyranny. The car pool dissolved, and both persons avoided Mercury.

During the progression, people were married, were friends or lived together. They reached a variety of depth in contact based primarily on the effort or decision of Mercury.

Mercury Semi-sextile Pluto

Mercury has a vivid spark of interest in cooperating with Pluto for the general good. This can develop when both persons extend their effort. If coercion enters the contact it tends to dissipate.

In the examples these people usually met in group situations. Their personal contact often implied the subtle touch of coercion. The hidden levels were not conducive to a firm contact, unless stronger aspects gave assistance.

The romances leaned to slow disintegration. Several marriages held a cooperative note in their common goals.

The progressions implied no more than a spark of effort put into any depth of contact.

Mercury Semi-square Pluto

Mercury has intermittent periods of tapping the heights and the depths of Pluto. He does not fully understand Pluto's position or attitude, so their cooperation vacillates. When Mercury has worry and anxiety, he easily becomes critical or irritated. With firm patience, he can endure the changes, as the condition will right itself or lose importance.

There are several drastic examples in an exchange of coercion. The covert depths of Mercury and Pluto may meet and clash.

A patient had Mercury semi-square Pluto of his psychiatrist.

A man with Mercury semi-square Pluto of his girl friend had a sublime relationship with her, until he used subtle force to begin a sexual relationship. He was indignant at her outrage and they separated.

A man with Mercury semi-square his friend's Pluto could understand his heights of integral aspiration, but not the depths of his decadent urges.

The Mercury semi-square Pluto progressions show a critical point in the contact when the heights or depths of the two can make or break the relationship.

Mercury Sesqui-square Pluto

Mercury has a sudden coercive situation or recurrent intervals of disruption in his cooperation with Pluto that tend to break up their present conditions.

Mercury may make the decision to begin or continue the contact or let the matter drop.

In several cases, the contact itself of the two people implied a coercive situation.

A man fell in love with a woman to whom his Mercury was sesqui-square Pluto. Their work and social groups were entirely diverse. Mercury knew that this woman would not be accepted into his church or his home. He rebelled at this social coercion. With four N parallels to the woman, he felt such a bond with her, that he broke many of his former life patterns to marry her.

In another example, two men were in the same cooperative group that broke up due to internal hidden conflicts. Mercury maintained his friendship with Pluto.

A romance in which the girl had Mercury sesqui-square Pluto of the young man broke up because of the intermittent periods of agitation when they could not seem to cooperate with each other.

In the progressions a woman had Mercury sesqui-square Pluto of the man when they began a love affair. Three years later his Mercury made the sesqui-square to her Pluto and they separated.

In the Mercury sesqui-square Pluto aspects, the two are able to develop a depth of relationship based on cooperation. Any element of coercion becomes agitating and disintegrative.

Mercury Inconjunct Pluto

Mercury is strongly attracted to cooperation or coercion with Pluto. He takes an initial interest and initiative in this. In time, these energies disperse into other areas. He can maintain a certain level of cooperation with Pluto.

Mercury's subtle force of will sometimes inverts, in these examples. He may begin his relationship with Pluto without preamble on a full cooperative basis that gradually disperses. An insidious coercion begins. When this continues, they separate, Usually it is stopped at a certain point, as Mercury wishes to continue the contact and to do so, he must cooperate.

The progressions show a climax of cooperation for the height of experience, or the depths of drastic force.

These heights were reached in the personal gain of romantic unity; or in the greater good of spiritual dedication that two people shared.

The depths of anguish occurred to Mercury when his brother Pluto was killed in a drastic accident. Mercury and Pluto both suffered shock when Mercury's husband, who was Pluto's father, was imprisoned for a criminal act.

Even when the energy of the inconjunct disperses, the impact remains.

Mercury North Parallel Pluto

Mercury has a different life pattern from Pluto in his heights of aspiration or covert depths. Their previous environment and experience have been diverse in these matters, with some few similarities.

They come together with spontaneous circumstantial ease on a matter of cooperation or coercion, that hits a responsive chord in both persons. Mercury will develop this as long as it fits in with his primary life pattern. His maximum expression of the general good, or his self gain, is in an area apart from Pluto.

The most intense examples of cooperation or coercion are sometimes included in the same case, where they invert.

A woman with Mercury N parallel Pluto to her brother-in-law joined with him in full cooperation to commemorate the memory of her husband, his brother, who had been the victim of the drastic criminal act of murder.

In another example, a woman had Mercury N parallel the Pluto of a man. They felt compelled to cooperative unity at any and all cost. (Refer to Mercury N parallel Jupiter). They ran away to Mexico to get quick divorces, so they could marry. The hidden areas in their relationship became coersive when the man was jailed, and later released under psychiatric care. Mercury was eventually impelled to separation, for survival.

One man who criminally assaulted a child had Mercury N parallel the child's Pluto. The coercion of this drastic act was not a part of his basic pattern except in his dark hidden depths of covert self gain.

The majority of the examples are less drastic. There are much more subtle forms of both coercion and cooperation. Often Mercury is a strong willed person who cooperates with Pluto as long as it fits his pattern.

The progressions show the time when the bond is strongest. Mercury usually takes the stand, "I will cooperate with you willingly, as long as it's on my terms." This is not always easy, as Pluto had dynamic covert areas of his own.

When both persons have a common level of spiritual aspiration, their cooperation can extend beyond personal benefit to the general good.

When they cooperate with the concealed will to power, they hit the responsive chord of rule or ruin in each other. In these examples the relationship is inverted; it turns inside out in a reversal of its former pattern.

In the examples where relationship was maintained, Mercury considered not only Pluto's benefit, but the integral general good. The Mercury N parallel Pluto is separative only to the extent, and in the area, that Mercury must fulfill his primary life pattern of aspiration for the greatest good in an area apart from Pluto.

Pluto is in North declination from 1864 to 1988. The definition for South declination will be included by formula. There is only one example chart with a Mercury S parallel Pluto. It is of Austrian royalty, in which Pluto was assassinated in 1898.

Mercury South Parallel Pluto

Mercury has a different life pattern from Pluto in his heights of aspiration or covert depths. They do have large areas of similarity that overlap.

They come together as a natural result of circumstances on a matter of cooperation or coercion. This matter implies a tension.

Though Mercury has a diversity with Pluto, their bond is such that he is impelled to an expression of the general good; or toward his self gain, with Pluto.

Mercury S parallel Pluto is unifying to the extent and in that area, that Mercury must fullfil a meaningful cooperative relationship with Pluto while maintaining his primary life pattern of aspiration.

Mercury Split Parallel Pluto

Mercury's maximum contact with Pluto is on a matter of the greatest general good, or on the deepest covert depths.

They have a level of cooperative or coercive experience apart from each other. They have areas of aspiration that are uninteresting or incomprehensible to each other.

They can have the greatest possible unity in their spiritual heights, and in their cooperation based on candor.

They tend to be forced apart in direct ratio to their covert, hidden or coercive depths. If there is a criminal factor in their separate lives, this can be an insidious force that comes between them.

In several of these examples there is the greatest cooperation for both spiritual aspiration and temporal benefit.

There are several examples of the student to the spiritual teacher (Guru) and of students who cooperate with each other for a unity of striving in the same teaching of ethos.

There are business cooperatives, and the semi-personal exchange of contact for a common good in work situations.

In two examples the maximum contact was reached in a coercive situation that inverted a former friendship into enmity.

The progressions have varying degrees of unity in ratio to the equality of the two in their covert as well as obvious areas.

They follow the same illustrative patterns as the natal examples, with the time of maximum contact specified.

Because these are two diverse lives that have come together in unity, there is some level of depth that each has apart from the other that is difficult to share. That separate area may be a covert, coercive or criminal experience, or it may be the ethos of spiritual aspiration.

The Chart Comparison on page 179 illustrates a group cooperative that influenced the culture of the time. They are four musicians in a world known rock group.

The Mercury and Pluto positions are the only ones recorded.

All four of these men have Mercury in aspect to the Pluto of the other three. They united in the closest cooperation for their mutual benefit and the general good. This illustrates well the coercive nature of Pluto, in that, without cooperation, there *is* no result of achievement. If any one puts his personal self gain above the good of the group, the unity is lost, and the result is disintegration.

The ethos of Pluto is unity in diversity. A musical group exemplifies this. There is the variety and versatility of the different instrumental and human voices, that combine for a common effect.

Chart Comparison Illustrating a Group Cooperative

Ringo 1961	☿ 5 ♌ 13 16 N.42 ☿ 27 ♋ 22 15 N.54	♇ 4 ♌ 00 23 N.15	☿ ☌ ♇
John 1961	☿ 7 ♏ 35 15 S.50 ☿ 27 ♏ 51 22 S.27	♇ 4 ♌ 06 23 N.20	☿p SPLIT ♇ ♇
George 1961	☿ 9 ♒ 56 18 S.31 ☿ 6 ♓ 00 11 S.10	♇ 4 ♌ 48 24 N.00	☿p ⚻ ♇ 1960
Paul 1961	☿ 18 ♊ 34 18 N.48 ☿ 23 ♊ 04 20 N.22	♇ 4 ♌ 02 23 N.48	☿ ∠ ♇

The essential dignities of the planets are of technical interest. Pluto is in its exaltation. Ringo has Mercury in its fall, but in the most vital aspect of conjunction. John has Mercury in its sign of harmony and in mutual reception to Pluto. George has Mercury in its exaltation; Paul has Mercury in its home sign.

Three of the men have Mercury in a T-square; Paul has Mercury trine, semi-square and sesqui-square the other three Mercury positions.

The Mercury to Pluto aspects between the four charts indicate the individual levels and reactions of cooperation to their group effort. The Mercury to Mercury discordant aspects indicate the communication conflicts that contributed to their eventual disbanding.

CHAPTER XIV

Mercury in Aspect to MC

THE M.C. RULES THE PUBLIC POSITION. It indicates the career and business, the reputation and credit, the honor and publicity. Mercury makes contact with the M.C. on these matters. There is an exchange in a public situation, or on a matter that receives credit or discredit on a level of public notice for both persons.

The Mercury-M.C. parallel marks the similarities and differences of the two people within their life patterns of business, career and reputation.

Mercury Conjunct M.C.

Mercury has a familiarity with the public position of the M.C. He understands or identifies from his own experience with M.C.'s reputation or business.

They come together to share, or experience at the same time, a matter of credit or discredit. As the conjunction is neutral, Mercury can acknowledge M.C. impartially.

Their public notice can be together in a common situation such as their careers or work together. Because of the impartiality of the conjunction, their *personal* reputations may differ. Some of these examples are prestigious, such as that of the president and vice president of a nation.

There are several examples of a wife with Mercury conjunct her husband's M.C., where she joins him in public situations relative to their position in life. That position can be national recognition or notoriety, or acknowledgement in their community circle.

One example is of two hippy boys who had a common reputation for long hair, beads and bare feet. Their public association together was in their musical group.

Two other men with Mercury conjunct M.C. had a brief association in which they both had discredit at the same time for a wild party that the police broke up.

The progressions have varying examples of public attention. A child with Mercury conjunct her mother's M.C. went with her on an international trip. Their normal household association was changed into the situation of meeting many people in public situations.

In other cases the two met in work and business contacts that were more or less prestigious.

Mercury Opposition M.C.

Mercury has something in common with M.C. of a public nature. Where they reach an understanding or identification on this matter, the full expression of their separate reputations do not blend. When a larger adaptation is required, the two are not given the same credit or discredit.

The examples generally imply some discredit. The two people often have their names linked together at a certain time in the public viewpoint, but do not have a similar reputation.

Two women with Mercury opposition M.C. were linked in newspaper publicity when Mercury's husband molested M.C.'s daughter.

In a marriage, a man and woman were accredited with their similarity of public work in the theatre, but had bad publicity with their divorce.

In some examples, there was a clash and separation when Mercury disapproved of M.C.'s reputation.

In the progressions a child and parent had Mercury opposition M.C. at the time when the parents divorced. There was no scandal, but the child felt keenly that this was dishonorable. She said, "Who would ever think that this would happen to me! How could *my* parents get divorced?"

Two women with Mercury opposition M.C. had a certain reputation in common in that Mercury married the man to whom M.C. had been married before.

In other examples there was community disapproval of the two people for the association, such as "What does she see in him? There'll be no good come of that match."

In other cases, the nature of their public work and business brings a period of separation.

Mercury Square M.C.

Mercury is strongly drawn to the image that M.C. represents. There are obstacles to making their contact public, or else an incompatibility in their reputations.

When Mercury acts with honor, he can build a reputable association with M.C. When Mercury has any possibility of dishonor, they are both in danger of public discredit.

The honorable situation with obstacles is illustrated by a movie producer with Mercury square the M.C. of his director. Their production was handicapped by limited funds, inadequate sets and crew. The finished product received little acclaim at the cost of a great deal of effort.

In some cases Mercury overrated M.C. in a way that caused a strain to their relationship. In other cases Mercury underrated M.C. in a conflict of wanting his own reputation to get first notice. Often there was a criticism or discredit between them.

In several examples Mercury lost contact with M.C. by presenting a note of dishonorable conduct, or by fear that their association would hurt his own reputation.

Some of the examples are distinctly a threat to the reputation, as in secret love affairs, use of drugs, or business dealings that could not stand too much investigation.

Though the potential threat is there, the Mercury square M.C. is more indicative of difficulty in getting public notice or credit of any kind.

The progressions have one case of an actress with Mercury square the M.C. of her mother. The actress received national acclaim for a movie role at the same time that she and her mother received discredit for their personal relationship.

There are other examples of relationships that, fortunately, received no public acclaim. If they had, one or both of the persons would have been ruined.

Mercury Trine M.C.

Mercury's contact with M.C. easily and spontaneously receives public notice with little or no effort.

Mercury spontaneously falls into contact with M.C. in a public situation or one that receives public notice, relative to their position in life. Circumstances and environment are such as to require little or no effort on Mercury's part.

When Mercury honorably enhances the M.C. he shares a full beneficial credit.

When Mercury is unappreciative or indifferent he may lose the benefit entirely.

In a classic example a woman had Mercury trine her son's M.C. She always related to his public activities with respectful admiration. She was proud of his scholastic acumen, his war record, and his business prestige. When they were together, he treated her like a queen, in attention and deference. In public situations they were a delight to watch for their pride in each other.

Among their acquaintances, Mercury was accredited with having brought up a fine son; M.C. was pictured in the newspaper with his charming mother.

In other examples "everyone" talks about Mercury and M.C. in varying degrees of credit or discredit, according to their relationship and separate conduct. "They make such a nice couple." "Did you hear about Mercury going into business with M.C.? It was reported in the Sunday paper." "I hear that Mercury's son M.C. is going to college on a scholarship." "Didn't Mercury do a beautiful job of redecorating M.C.'s showroom? I hear that his prices are high." "Do you think Mercury will stay with M.C. after the way he handled that deal?"

These examples for the most part enhance the reputation favorably. In a few cases the publicity that came so easily was for distressing reasons.

A woman had Mercury trine the M.C. of a fine son who was killed in foreign battle. A man with Mercury trine the M.C. of a woman attacked her and was jailed with discredit and scandal.

In one case two men received publicity for a business failure. Mercury was credited with bad judgment and M.C. with incompetence

In a few cases Mercury was either totally indifferent to the M.C. and his reputation, or avoided any public notice. These were minor, or short term, contacts.

Mercury Sextile M.C.

Mercury has the ability and opportunity to publicly acknowledge M.C. for his career or personal conduct. There is a certain time when the two are drawn together easily in a situation where they can build credit or discredit.

Mercury has the choice of whether to make this matter public or not.

Where there is a respectful and honorable exchange Mercury will speak highly of M.C. relative to his position in life. He may tell everyone he knows, or he may, literally, broadcast it on the radio.

Mercury's conduct in some cases makes the public announcement, as in the example when a man and wife were separating, and Mercury shot himself. They both had bad publicity.

In another marital separation the man put the notice in the paper, "I am no longer responsible for my wife's debts."

Usually the aspect is harmonious. Mercury will speak well of the accreditable conduct of M.C. and omit the areas that he does not care to make public.

In some cases Mercury makes a firm choice of keeping the whole affair secret, as he would discredit himself. In all cases Mercury has the choice of just how much, or how little of his contact with M.C. will be publicly known.

The progressions show specific times when Mercury and M.C. are in a public situation together, or when there is a personal exchange that could give them a reputation.

A mother had Mercury sextile the M.C. of her son when he attained a community post for which she had helped him campaign.

A man had Mercury sextile the M.C. of a business acquaintance whom he cheated.

A husband had Mercury sextile the M.C. of his wife for a five year period after he was released from jail. He reestablished his credit and honor at this time.

A woman with Mercury sextile the M.C. of a friend, defended his bad reputation. Another woman with Mercury sextile the M.C. of a friend discredited him.

Several cases of secret affairs remained secret, as Mercury only made public as much as he chose.

Mercury Semi-sextile M.C.

Mercury has a vivid spark of interest in the career, public position, or reputation of the M.C. If both persons put the effort into developing an honorable contact, this can be prestigious.

When other aspects add substance and endurance, the two people can use this spark of common interest to develop a growing relationship.

In a few cases the people met in work or career situations in which Mercury was able to give M.C. public acknowledgement and credit.

In one example Mercury was quite concerned about M.C.'s reputation, and made an effort to advise M.C. to be careful of a certain situation. M.C. blythely continued to flaunt bad judgment, and Mercury avoided further contact.

In the progressions Mercury both openly praised, and privately discredited M.C. in different cases. Sometimes these two reactions occured in the same situation.

Any discredit generally remained limited to a small area of publicity. In a few examples the better reputation was widely known, relative to their position in life.

Mercury Semi-square M.C.

In Mercury's contact with M.C. he had intermittent periods of worry and anxiety about the M.C.'s reputation or public position, that easily causes criticism or irritation.

Mercury does not fully understand M.C.'s position or attitude in this matter. He can handle these changes with firm patience and conditions will right themselves or lose importance in time.

In one example two girls had Mercury semi-square M.C. who were friends for many years. Mercury worried about M.C.'s popularity in school, as M.C. had the reputation of being flirtatious and flippant. Later Mercury could not understand why M.C. chose to marry a certain boy, and was critical of her decision. Mercury was often irritated when M.C. broke public appointments. Their friendship did endure through these vacillating periods.

In many cases Mercury's worry included concern for his own reputation, as a result of M.C.'s conduct.

Mercury Sesqui-square M.C.

Mercury has a sudden situation with M.C. or recurrent intervals of disruption that tend to break up their present public position or reputation. Mercury may make the decision of whether to begin or develop the contact, or let it drop.

A long-term contact was of a man with Mercury sesqui-square the M.C. of a woman. They were both in the theatrical field, a career that gave them prestige and publicity through the years.

When they met, on their first movie together, they were linked romantically, despite the fact the M.C. was married. They survived this initial agitation and built a friendship. At various times, both together and separately, they had bad publicity and public discredit, as well as international acclaim.

In every case there is some period of agitation for both, over the public favor or disfavor they receive.

One romance broke up because the woman, Mercury, had a bad moral reputation, and M.C. felt that he could not jeopardize his professional career with their affair.

Another man and woman moved into an apartment with the open declaration of their intent to live together; if someone didn't like it, that was their problem. The agitation occurred when the girl's mother came to visit.

In one progression case, Mercury had a sudden change of reputation to prestige and acclaim that broke up his existing conditions. He and his wife, M.C. moved into a different social and economic circle.

Mercury Inconjunct M.C.

Mercury is strongly attracted to the public image or reputation of M.C. He takes an initial interest and effort to develop their career or position. In time, Mercury's attention disperses into his own public field. He can retain a respectful relationship to M.C.

In these examples, M.C. represents some image of prestige or capacity to Mercury.

A woman Mercury fell in love with M.C. a young man from an important family. After their marriage she went back to her own personal career, as well as maintaining a supportive image to his public life. They were a credit to each other and highly spoken of in their community.

In several other cases Mercury carried the prestige and business power, but gave credit and publicity to M.C. for a certain job or ability.

The progressions are more indicative at a specific time when Mercury is drawn to M.C. on a matter of public position or interest.

A woman with Mercury inconjunct the M.C. of a man worked for him during this time. She initially gave him enthusiastic credit, but this quickly dispersed when she found his methods not to her idea of honorable credit.

A young woman had Mercury inconjunct the M.C. of her mother when her father was killed. The death had a great deal of publicity because of the father's public position.

Mercury North Parallel M.C.

Mercury has a different life pattern than M.C. in his career or business, reputation or publicity. Their previous public experiences have been diverse, with some few similarities. They come together with spontaneous circumstantial ease on a matter in which they are linked together publicly. They respond to each other in this matter. Mercury will develop this as long as it fits in with his career or reputation.

In most of the examples the two people have some bond in common in their public image. There are entertainers of different types, hippy kids, people whose business affairs come together, people who have a family connection. Their basic career patterns and reputations are diverse.

There was a long-term friendship between an artist, Mercury, and an actress, M.C. Mercury received public acclaim for his painting of M.C., as she had a large area of prestige.

In other examples, the two maintain association only for the length of time that they spend in the same business or career.

Two brothers with Mercury N parallel M.C. went into diverse fields. One took an army career, the other became a school teacher. They had some similarity of reputation during their school years when they both had athletic honors.

In one example, the two came together at a time of scandal and disrepute.

The progressions show good, bad, and indifferent reputations relative to the position in life, and the conduct of the two people. It is a public matter in each case. There are marriages and divorces, deaths, career prestige or discredit, or changing career patterns.

The Mercury N parallel M.C. is separative to the extent, and in that area, that Mercury must fulfill his primary public position in an area apart from M.C.

Mercury South Parallel M.C.

Mercury has a different life pattern than M.C. in his career or business, reputation or publicity. They have large areas of similarity that overlap.

They come together as a natural result of circumstances where they are linked together publicly. The matter implies a tension. Despite their diversities, the bond is such that Mercury is impelled to develop their public contact.

A typical example is the case of two brothers with Mercury S parallel M.C. They both had scholastic honors and a reputation for intellectual acumen.

Mercury went into business, M.C. became a lawyer. Their diversity was in their basic goals and type of reputation. Mercury found that he needed legal counsel, and their public work often was together.

These people usually had long periods of similar work or career. Often their personal reputations overlapped in the same way, usually for credits.

The progressions are more specific of a time when Mercury is concerned or adds influence to M.C.'s reputation or public position.

In one case Mercury demanded that M.C. change his conduct in order that they both avoid getting a bad reputation.

In another example a man with world-wide prestige had Mercury S parallel the M.C. of another highly reputable man with whom he began a business partnership.

In several cases Mercury's reputation was such as to cause discredit to M.C.

In every example there is some time or some measure of tension between them that requires a positive, active contact.

Mercury S parallel M.C. is unifying to the extent and in that area that Mercury must fulfill a meaningful public relationship with M.C. while maintaining his own career and reputation.

Mercury Split Parallel M.C.

Mercury's maximum contact with M.C. is in a public position. Each has a level of reputation and public experience apart from the other. They have areas that are uninteresting or incomprehensible to each other.

They can have a cohesive unity in candor of mutual honor and respect. They tend to be forced apart by any discrediting conduct, or hidden factors, that come from the separate areas between them.

Where this aspect does not preclude personal relationships, it markedly indicates that the maximum extent of the relationship is a matter of public record.

Two friends with Mercury split parallel M.C. were involved in a business scandal at their place of work. Neither were directly involved to the extent that their reputations suffered, but it was an impact from which they separated into different jobs.

The examples can be completely opposite in quality and still illustrate the maximum contact as being public. Two women with Mercury split parallel M.C. gave each other honor, credit, and esteem. Their contact was both in the home and on the job at intermittent periods. They had no occasion or inclination to discredit each other and always spoke in estimable terms.

Two other women with Mercury split parallel M.C. had a neighborhood contact. Mercury discredited herself by gossip and slander of M.C.

In the progressions there is a marriage and a divorce. The other examples show a time when the two work together or meet in more or less public situations.

Mercury shares his prestige with M.C. or accredits M.C.

In the majority of cases, the two people have a diverse business or career. In the few examples where they have the same career, their reputation and publicity vary widely.

The Chart Comparison on page 192 illustrates two political figures who both were assassinated in public situations.

These two men lived 100 years apart in time. With the Mercury-M.C. aspects they had a similarity of reputation in that one was a political candidate for Presidency, the other served as President of his country for four months. One was shot in the head by a fanatic in a hotel kitchen; the other was shot in the back by a disappointed office-seeker in a railroad station. The one was born with advantages of wealth and education; the other was born in a log cabin and supported himself throughout his schooling.

Both were men of great charm; intelligent, well spoken, and able to see the issues of their day clearly. Neither were allowed to fully express or put their beliefs into practice, because of having their lives and careers cut short.

The Mercury opposition M.C. shows the similarity and diversity in their reputations, as well as the publicity that both had for the harsh shock of public violence.

It may be recommended to the astrologer interested in political charts that he use the Mercury Method of chart comparison with the presidents of the United States to each other and to the chart of the United States. In this study many valuable indications are found.

Chart Comparison Illustrating Two Political Figures

Natal	RFK to JAG	Natal	JAG to RFK
☿ 19 ♐ 38 25 S.35	⚻ ☽ 1°	☿ 29 ♏ 45 12 S.00	⚹ ☽ 3°
	∠ ♂ 4°		S.♇ ♂
	⚹ ♃ 4°		☌ ☉ 2°
	□ ♄ 5°		☌ ♄ 10°
	☍ MC 8°		△ ♅ 8°
			⚼ ♇ 1°
			☌ MC 6°
			⚹ Asc 1°

CHAPTER XV

Mercury in Aspect to ASC

THE ASC RULES THE BODY AND PERSONALITY, the disposition and attitude. It is the temperament, viewpoint and personal prejudice.

Mercury makes contact with Asc on these matters. There is an intimate exchange of affinity or antipathy in their personal affairs.

The Mercury-Asc parallel is a contrast of similarities and differences of two separate persons, in their individual life patterns.

Mercury Conjunct Asc

Mercury has a familiarity with the personality of Asc. He understands or identifies from his own nature, with the disposition and attitude of Asc.

They come together to share, or experience at the same time, events or conditions that shape their intimate affairs.

As the conjunction is neutral, the contact can be made in affinity or antipathy. However the position of Asc is highly personal so Mercury's acknowledgement is not entirely impartial. There are several short-term contacts where circumstances and other aspects were prohibitive to intimacy but even in these, both persons reacted to awareness of the other.

Often there is body contact, ranging from the intimacy of birth, to the sexual exchange in marriage and affairs, to the embrace of friend or kin.

Mercury identifies with the Asc more intimately than with any of the other planets, except Mercury conjunct Mercury. He finds some manner of expression, or appearance, or disposition in Asc that he has himself. Even when the contact does not publicly reveal any importance, the inner changes of attitude and personality of the two are marked in this close contact.

Their most intimate exchange may be together, or their contact may occur at a time when both are having physical or personality changes. In either event, Mercury can mark a milestone in his life by his contact with Asc.

Mercury Opposition Asc

Mercury has in common with Asc certain attitudes and inclinations. Where they reach an understanding and identification on a personal level, their basic personality, disposition and attitudes do not blend. They can maintain association at a certain level or at intermittent times. When the larger adaptation of a common environment is required, they build up inflexible differences.

In these examples we have a fraternal contact of ten years cohabitation, friendships and business acquaintances of longer periods. The aspect is not antipathetic, quite the opposite. Mercury is attracted to Asc. The extent, depth and substance of the two personalities are unequal. In the more shallow examples, Mercury reached no more than the surface image of Asc.

The inflexible impasse that the two develop is usually in that their personal affairs are either in conflict or do not interest each other.

The progressions have examples of friendship on an intermittent basis of contact. Mercury gets along well with Asc but not for long.

Two love affairs were begun at this time. With one, the aspect moved out and the couple married. The other was sheer disaster for them both. Their attitudes clashed so violently in the intimate contact that both withdrew in a state of shock.

In a long-term relationship, the progressed Mercury opposition Asc shows a period of separation in viewpoint or personal or physical experience.

Mercury Square Asc

Mercury is strongly drawn to the personality of Asc, but there is an obstacle of incompatibility in disposition and attitude.

When Mercury makes the effort to understand Asc he can build a constructive and intimate exchange of affinity.

When Mercury has prejudicial attitudes and inflexible habits he may find nothing of value in the contact.

In some cases Mercury had such respect and affinity for Asc that he broadened his own viewpoint in their development of exchange. The give and take of two distinct personalities enriched them both. We have this illustrated with a student and teacher, a mother and daughter, with business associates and friends.

In other examples, Mercury established a personality conflict with Asc that led to a crisis. Both lovers, friends, and enemies clashed in their personal viewpoints and prejudice until they separated.

The progressions hold some measure of conflict that is not too severe. It is more the nature of a clash of some viewpoint in temperament or attitude. Mercury can handle this easily with an effort to adaptation.

Mercury Trine Asc

Mercury has a spontaneous contact with Asc in which there is a natural affinity. Circumstances and environment are such as to require little effort on Mercury's part, to develop a relationship.

When he makes no effort to enlargement or depth he may lose the contact entirely. When Mercury pursues the opportunity he can build an exchange of personal benefit.

The Mercury trine Asc by itself is lacking in depth, and connotes no more than a pleasant personality exchange. When other aspects add interest and effort, we have a more intimate relationship, of every type – friends, lovers, family and business contacts.

The progressions bring the people together easily in natural circumstance. Mercury responds easily to the personality of Asc. They share some temperament, viewpoint, disposition, or attitude.

Mercury seldom does much more than is required, or that which comes easily.

Mercury Sextile Asc

Mercury has the ability to develop a personal relationship with Asc. There is a certain time when the two come together easily and the opportunity is presented to develop an intimate contact, physically or in attitude.

If Mercury makes neither effort nor response the time passes and the contact is lost.

The two come together in the home, at work, through friends in common, or because of similar interests. Their temperament, viewpoint, and attitudes can blend.

In every case there is some exchange of affinity. At the least this is the pleasant chat over morning coffee between two neighbors. In other examples, the two began love affairs, deep friendships, marriage, or cohabitation.

In one of the progressions, there was a business exchange that was highly personal only in the sense that it hit the disposition and prejudice level.

This revealed the attitude of the two people to each other during their short time of contact.

Mercury Semi-sextile Asc

Mercury has a vivid spark of interest in the personality of Asc. This can grow into an intimate relationship if both persons make the effort to develop it.

In one example a man had Mercury semi-sextile the Asc of a woman friend. Periodically he would make an effort to begin a romantic relationship. With Mercury opposition his Asc, the woman would respond in affinity, but their deep attitudes and temperament would clash, and they would separate with a certain bewilderment. Though they liked each other, they just couldn't seem to get together in a depth of contact.

In some examples the spark of personal interest did lead to friendship, love affairs, and marriage.

When other aspects did not add substance, the Mercury semi-sextile the Asc went no further than the stimulation of interest.

Mercury Semi-square Asc

In Mercury's contact with Asc, he had intermittent periods of worry and anxiety that easily became critical or irritable. Mercury does not fully understand Asc's viewpoint and attitude.

Mercury can handle these slight personality changes with firmness and patience, as their personal affairs will straighten out, or lose importance.

In each case, Mercury had some concern about the attitude, personal appearance, health or disposition, or personality problems of Asc.

Asc often displayed temperament, prejudice or a vacillation in viewpoint.

One man, Mercury, worried because the woman, Asc, was in love with him. Another man with Mercury semi-square a woman's Asc was concerned because she couldn't make up her mind between him and his rival.

The various relationships endured through these changes, unless other factors were prohibitive.

Mercury Sesqui-square Asc

Mercury has a sudden situation or recurrent intervals of disruption with Asc that tend to break up their present relationship.

Mercury can make the decision of whether to begin or continue their contact on the new lines presented, or whether to let the matter drop.

The long-term contacts had periods of agitation because of unexpected difficulties in the environment, or in changes that one or the other would make. One couple were broken up by gossip and the intervention of other people.

Another couple with Mercury sesqui-square Asc had the agitation of a divorce, after which they decided to remarry.

Two men with Mercury sesqui-square Asc worked together under conditions of personal temperament and erratic disposition. In the progressions, friendships were begun and ended, with agitation and disruption.

A young woman with Mercury sesqui-square the Asc of her father was personally upset when he decided to remarry.

The sesqui-square is indeterminate in nature and in timing.

Mercury Inconjunct Asc

Mercury is strongly attracted to the personality of Asc. He takes an initiative in developing their contact. In time, this disperses but he can retain a personal relationship in affinity.

In all of the examples Mercury had an immediate affinity for Asc. In a few cases, this dispersed quickly when he became better acquainted with the attitudes of Asc. Usually it was maintained and developed into a highly personal exchange.

There were marriages, love affairs, friendships and familial devotion.

In the progressions, Mercury has a climax of interest in the Asc, as his attitude, body, or personality are important to Mercury. A young boy had Mercury inconjunct his mother's Asc when mother became pregnant. The boy was fascinated by the change in mother's shape, as well as the promise of a brother. By the time the baby came, his interest had dispersed into a mild acknowledgement.

In other examples Mercury is strongly drawn to Asc with some solicitude or attraction. In several cases, the two supported each other's attitude during a period of tension. There was a death in the family, Asc was divorced, or they moved. Each case brought Mercury and Asc closer during this period.

Mercury N Parallel Asc

Mercury has different personality patterns than Asc. Their physical experiences and environment have been diverse, with some few similarities.

They come together with spontaneous circumstantial ease in a personal attraction to which they both respond. Mercury will develop this as long as it fits in with his primary attitudes.

These people are attracted together but their personal diversities blend no better than the opposition. There are long-term friendships, and family contacts, but even in these Mercury spends a greater amount of time in an environment that fits his own disposition, than he does with Asc.

The progressions show people being drawn together, sometimes in affinity and devotion that endures in intermittent contact. One marriage began at this time, as the aspect moved out. Generally they can only stay in the same environment for short periods of time. The temperament and viewpoint of their separate attitudes tend to clash.

The Mercury N parallel Asc is separative to the extent and in that area that Mercury must fulfill his personality patterns and personal environment in an area apart from Asc.

Mercury South Parallel Asc

Mercury has a different personality pattern than Asc but many of their attitudes overlap in similarity.

They come together as a natural result of circumstances in a personal contact. There is a tension that impells Mercury to respond to Asc.

Mercury may feel that he is drawn to Asc in spite of himself. He declares that he has his own interests and personal affairs to take care of, but the tension of attraction is impelling.

In one long-term contact between a mother with Mercury S parallel her daughter's Asc, she would periodically respond to Asc at times of tension in joy and sorrow. There were the familial bonds of birth and death, problems and successes.

In one progression, a girl had Mercury S parallel the Asc of her boyfriend's mother. When the boyfriend was in a serious accident, Mercury went to the mother's side to comfort and support her. They found a great deal of similarity in attitude and temperament that they appreciated in each other.

The Mercury S parallel Asc is unifying to the extent and in that area that Mercury must fulfill a personal relationship with Asc while maintaining his own pattern of environment and basic attitudes.

Mercury Split Parallel Asc

Mercury's maximum contact with Asc is on a highly personal matter. Each has a level of attitude and environment apart from the other. They have viewpoints that are uninteresting or incomprehensible to the other.

They can have an intimate unity in ratio to their equality of temperament and attitude. They tend to be forced apart by the incommensurate facets of their separate personalities, or by any covert or hidden factors.

The complexity of the personality is illustrated by a man who had Mercury split parallel Asc of two of his friends. Mercury's personality was not similar to that of either of these friends. Mercury identified with one friend in their similar viewpoints on books, entertainment and business. He identified with the other friend in their similar viewpoint on morality and conduct. Mercury had an intimate exchange with both friends while maintaining his own individuality.

Usually there is a wide open exchange, that insures a maximum personal contact of friendship, marriage, romance, and work relationships.

Even in the examples when Mercury concealed his motives and attitudes there was a personal reaction between them, but one that was separative. Where there was an inexplicable antipathy, there was sure to be hidden viewpoints that were withheld.

Because the maximum contact was in personal attitude, the Mercury split parallel Asc does not indicate events.

This Chart Comparison illustrates enmity. It is of two men named Bill and Charlie, who worked and lived together.

Natal	Bill to Charlie	Natal	Charlie to Bill
☿ 18 ♑ 05 8 S.20	⚹ ☽ 4° ⚹ ♄ 4° □ P 1° ⚹ MC 2° □ Asc 9°	☿ 5 ♏ 42 17 S.50	∠ ☉ 1° □ ♀ 2° ☌ ♂ 4° ☍ ♃ 5° □ Asc 6°

Bill was a Sagittarian, with a good humored pleasant personality. He worked as a machinist.

Charlie was a Scorpio, quiet, retiring, and hard working in his job as a welder. He was going to school nights to study engineering.

When they met, Charlie's Mercury was conjunct Bill's Mercury. Bill came to work at the shop where Charlie had been employed for 6 years. They had a familiarity in their work and an easy communication on the identical experiences that they shared during the next 6 months.

(Mercury sextile Saturn, natal.) Bill not only accepted the responsibility of working with Charlie, but he respected Charlie's effort of improvement in going to night school. Charlie's finances were very limited at this time.

(Mercury sextile Moon, natal.) Bill responded with trust and sympathy to Charlie. He offered Charlie a room at his home.

(Mercury sextile M.C. natal.) Bill spoke highly of Charlie to his wife and family. All the boys at work knew that the two men were friends.

(Mercury semi-square Sun, natal.) Charlie related to Bill, but with some concern. He was accustomed to a retiring life, and occasionally Bill's wide open buoyancy irritated him.

(Mercury square Venus, natal.) Charlie couldn't help but admire Bill's charm. He enjoyed the easy warmth that Bill gave so easily and naturally to everyone.

Charlie moved into a spare room in Bill's home. He liked the family, but an ambivalence of temperament began to build up. When Charlie got home from work, and later from school, he wanted quiet to rest and study. The household was noisy and active. Bill liked to play poker and drink beer and have a few friends over in the evening, or turn on the TV loudly.

(Mercury conjunct Mars, natal.) Both men were hard working and skilled on the job. Shortly after they met, Bill caught his hand in a machine that mangled two of his fingers. He lost a month's work but kept the use of the hand.

Charlie approached his labors with a steady industry. He could acknowledge Bill's more demonstrative display of effort without feeling any need to compete.

(Mercury opposition Jupiter, natal.) Taking a room with Bill was a big help to Charlie. He had a ride to and from work and his meals provided. The expense was less than when he lived by himself and he appreciated this. However Charlie's nature was not demonstrative. His area of expansion was in his studies to self-improvement.

Bill's urge to expansion and pleasure was in sports, cards and friends, and easy living. He made good money and spent it so quickly that he was usually in debt. He drove a big car. Charlie was not in agreement with this philosophy.

(Mercury square Pluto, natal.) Bill had been strongly drawn to cooperate with Charlie for their mutual good at work. They both attended Union meetings and were in tight agreement on the issues that came up.

In living together, the incompatibility of their deeper aspirations became pronounced.

Bill often put Charlie in the position of having to do an action. "Hey, Charlie, pick up some beer on the way home from school, will you?" "Charlie, will you get the car lube'd today? I'm going on a fishing trip with Sally and the kids in the station wagon." "Charlie, you'd better pick up the parts we need or we won't get the order out from work tomorrow."

Charlie resented this, but said nothing.

(Mercury square Asc, both natal.) The incompatibility of their dispositions and attitudes accumulated.

Neither man seemed able to break through to an understanding communication of those areas of diverse viewpoint.

The personality conflict led to a crisis, first in short outbursts of temper, and finally with a raging battle.

Charlie had Mercury progressed to square Bill's Saturn. Bill decided to paint the outside of his house. He informed Charlie that they'd get started with the west side on Saturday, the day that Charlie was studying for a final exam.

Charlie exploded. He told Bill what he could do with his paint and his house, and he moved out.

The men parted with hard feelings and enmity. They continued to work together but with sparse civility.

CHAPTER XVI

Conclusion

THE *TIMING* OF A SITUATION or event in human relationship must be evaluated with the study of the separate charts as well as the compared charts. Generally, when a Mercury to planet aspect in chart comparison reaches its exact degree by progression, the event or situation reaches its peak. However, if neither person has *that planet* in aspect in their personal charts, the Mercury to planet indication will be effective only as an attitude of exchange.

Consider, as an example, that Mary has Mercury square John's Mars by progression, reaching the exact square in February 1970. How will this express and *when*? Do they work together, are they constructive in effort, competitive, hostile, sexually attracted? Their individual age, situation and former patterns will set a criterion of judgment on these matters.

Does Mary have a major progressed aspect to Mars in her chart? Does John? Is either one of them accident prone or facing the possibility of surgery? Are they already involved in an action situation or a sexual relationship?

If there is no Mars progression in either chart, the Mercury square Mars by chart comparison does not indicate trauma, violence or harsh discord. It is indicative of obstacles in their attraction to a common effort of energy; or a period of antagonism. If either one has a major progressed aspect to Mars in their personal horoscope, *that one* will be most inclined to express the aggressive energy in the relationship. As Mary's Mercury reaches the exact square to John's Mars in February 1970, the tendency to strife, effort and increased energy reaches its peak at that time.

Consider another example of Mary with Mercury trine John's Jupiter natally. To what extent may she give benefit to John, and how may he return her faith?

Does either chart indicate wealth? Do they have business dealings? Is there a sport or adventure in their environmental patterns? May they have travel, education or religion in common that broadens their scope of experience?

Their separate lives and individual charts will define the specific reaction of the chart comparison indication. Does one or both of them have a major progressed aspect to Jupiter in their separate charts? If so, the one to whom the benefit will be greater as well as the type of benefit is indicated by that progression.

In the Mercury Method of Chart Comparison, Mercury aspects to the Sun, Moon, Venus and Asc are indicative of inter-personal relationships of family, friends and lovers, of cohabitation and physical contact.

The Mercury aspects to Sun, Mars and Saturn are more applicable to interaction on the level of goals, effort and security.

The Mercury aspects to Saturn and Jupiter, Mars and M.C. pertain to business exchange, financial loss and gain, mundane competition and public affairs.

The higher octave planets, Uranus, Neptune, and Pluto, have a much greater pertinence to the individual than to formulae. They are more delicate, subtle and complex to read. When it is said of Uranus that there is a unique quality of changing patterns, this must of necessity be geared to fit the individual. To the traditional little old lady from Pasadena, a trip to Seattle to visit her eccentric son may be electrifying and an unusual experience of change that alters her normal patterns radically. To a vivid, active person with marked areas of independence, a Mercury to Uranus aspect will indicate an exchange of much greater dynamics and impact.

The subtlety of Neptune may be lost completely to the prosaic mundane personality. Even to the sensitive and imaginative person the Neptune exchange can remain an aesthetic experience, or one that is not revealed. Neptune is never obvious. Even in its most coarse expression of deceit or scandal, it implies hidden factors. In its most sublime form of religious ecstasy, the experience is subjective.

As Pluto connotes the heights and depths of human experience, its effect will only be recognized by the individual to the extent that he has himself plumbed his recesses and subliminities.

Both Neptune and Pluto are to be read with the greatest caution. In many cases, only the two individuals involved will be able to comprehend the full implications of the intimate experience.

All three planets, Uranus, Neptune, and Pluto, sometimes seem to strike a dumb note. In reading the compared aspects of Mercury to Uranus, Neptune or Pluto, it is advisable to understate, unless the more drastic or sublime effect is obvious.

The complexity of the parallels often implies a problem to work out of interaction. Both North and South parallels indicate inhibiting circumstantial factors to overcome. This does not indicate that contact on these matters should be avoided. but that greater effort is required to gain a greater result than that of the other aspects.

With multiple N parallels, vivid changes in life style and circumstance are made by deliberate effort to optimal results. With multiple S parallels the necessity of response does accomplish a successful interaction.

With multiple split parallels, the two people have so much activity and experience apart from each other that a consistent enduring relationship is not indicated. A maximum contact is effective in the areas where they do cross each other's lives.

In the Mercury Method, all the aspects of Mercury parallel a planet show the most complex areas of interaction between two people, that can make or break their relationship.

In every aspect defined in the Mercury Method of Chart Comparison, the final interaction result is subject to the *choice of the individual.* The *pattern* is defined. *What Mercury does about it is up to him.*

The harmonious aspects can be disregarded, taken for granted, or accepted as casual benefit; or they can be easily carried through to happiness and productivity.

The discordant aspects can be negatively used as inflexible, hostile agitation and worry; or Mercury can make the necessary effort to overcome obstacles to gain effective results. *Every example repeats this in obvious terms of actual situations.*

We begin and conclude with the individual. The right of choice remains within the will of man. Astrology indicates possibilities and the percentage of probability. There is no inevitability.

The superior man is the individual who persists in using the discordant aspects of life in a constructive manner and the harmonious aspects of life with integrity. It is he who has the freedom of choice.

INDEX OF SPECIFIC ASPECTS